INTRODUCTION TO
ERGONOMICS

W. T. SINGLETON

Professor of Applied Psychology
University of Aston in Birmingham, England

WORLD HEALTH ORGANIZATION
GENEVA
1972

Reprinted 1978

INTRODUCTION TO ERGONOMICS

Contents

Preface

As long as the tools and machines used by man were simple it was possible to produce satisfactory designs by purely empirical methods. Now that machinery has increased enormously in complexity, however, the empirical approach no longer suffices. To ensure the maximum efficiency of operation, to minimize the possibilities of human error, to reduce fatigue, and to eliminate as far as possible any risk to the operator, it has become necessary for the designer to adopt a scientific approach based on anatomical, physiological, and psychological considerations of human capacities and limitations. Through the collaboration of specialists in these disciplines the new science of ergonomics has emerged and in the industrialized countries ergonomists are being trained in increasing numbers and national ergonomics societies are being formed. On the other hand, in countries now entering a phase of rapid industrial development, where there is an acute need for engineers, designers, administrators, and occupational health workers familiar with the principles and applications of ergonomics, scarcely any training facilities exist. Moreover, although some textbooks on the subject have been published, they are geared almost entirely to the needs of the developed countries.

It was with this situation in mind that WHO invited Professor W.T. Singleton to write a manual on ergonomics that would set out the basic principles in simple language with an abundance of illustrations and could be used as a teaching aid in the developing countries. Professor Singleton had already organized a WHO training course in ergonomics in Bombay in 1967 and, through his contacts with the participants and visits to factories in India, gained a first-hand knowledge of the special problems facing countries that are still in the early stages of industrial development.

A first draft of the manual was submitted for review to ergonomists and occupational health specialists in 10 countries (see page 45) and their comments were taken into consideration in producing the final version. To increase the value of the manual for teaching purposes, each chapter concludes with a summary of general principles and notes on special cases; the principles set out in the chapter are further exemplified in graphic form in a series of illustrations. Throughout the book, emphasis is placed on the practical aspects of the application of ergonomics.

Introduction

Ergonomics—the technology of work design—is based on the human biological sciences: anatomy, physiology, and psychology. In general terms, anatomy is concerned with the structure of the body (the size and the way it is constructed); physiology is concerned with the function of the body (the biological processes that maintain it); and psychology is concerned with behaviour (the adaptive responses of the organism to its environment). The border lines between these disciplines are not rigid: for example, functional anatomy, which is that part of general anatomy of main interest in ergonomics, could be regarded as physiology. Similarly, there are many adaptive mechanisms in the body such as those concerned with homoeostasis, that could be considered as within the scope of either phsysiology or psychology, and in fact are often called physiological psychology.

It may be objected that no one person can be fully competent in all these three sciences. In fact, only parts of them are relevant to ergonomics; unfortunately, it is precisely these parts that are probably given least attention in medical courses. In general, for both anatomy and physiology, the behaviour of the intact organism only is of interest. The micro-characteristics of muscle chemistry, for instance are not directly relevant and certainly the pathology of such biological systems is outside the frame of reference of ergonomics. In the case of psychology, interest is restricted to human psychology and also to healthy normal behaviour. The general approach is that employed by the experimental psychologists—viz., the evidence and concepts used are those that have arisen from data acquired by controlled experimentation.

In spite of these difficulties in separating areas of disciplines, some classification must be made if the subject is to be studied systematically. This book covers three topics of equal importance within ergonomics: the biology of work, the psychology of work, and methods of studying work.

The biology of work is described in Chapters 1-5. This section covers anatomical and physiological aspects and some concepts that are usually regarded as essentially psychological. The application of forces and the problems of size and posture are largely anatomical questions, while the generation of energy and problems of climate may be regarded as mainly within the domain of physiology. The study of the effects of light, noise and vibration requires some appreciation of physiology and psychology and also of the elementary physics of these phenomena.

The psychology of work is described in Chapters 6-10. This section deals with the study of behavioural aspects, which is mainly dependent on psychological knowledge, although some topics, such as control design, relate also to anatomy and others, such as shift work, have physiological and psychological components. From the point of view of the psychologist, man is an information-processing device concerned with making decisions based on a variety of inputs and communicating these decisions by a variety of outputs. In some respects his behaviour is analogous to that of a computer, but the analogy is not a close one since in the computer field there are no problems of arousal, fatigue and circadian rhythms. Because, unlike the computer, the human operator can function at many different levels of arousal, it is necessary to study the question of motivation, which in turn requires a study of financial and non-financial incentives and interactions between people. Variations in level of activity with time of day have relevance to problems of shift work, rest pauses, and length of working day. These are highly complex questions to which there are never simple answers, but since these problems do arise in every industrial organization, some attempt must be made to deal with them even though the solutions may be only approximate.

Methods of studying work are described in Chapters 11-15. A thorough knowledge of the biological sciences is not in itself adequate to ensure competent performance as an ergonomist. Ergonomics is a technology and by definition every technologist has a repertoire of specialized techniques which he is expected to understand and to utilize but which are not available to people without training in this specialty. In ergonomics these techniques centre around the conduct of studies involving the measurement of human performance. The expertise required in this field is different from that needed to do experiments in the physical sciences or to acquire clinical evidence about individuals. Thus, although an ergonomist may have been trained as a physician or an engineer, he cannot claim to be an ergonomist on either of these grounds alone. Since the evidence that is obtained from human performance studies is complex and multifactorial, its presentation is as important as its acquisition. All the relevant techniques are dealt with in Chapters 11-15. Again, because of the complexity of the problems, even the skilled ergonomist needs some guide to structure his thinking. This he can obtain partly from his understanding of the way in which the subject has developed and what it tries to cover, and partly from an appreciation of the general pattern of ergonomics studies in relation to the design of work — in other words, the common questions that arise in all work design and the order in which they occur. These topics are dealt with in the final chapter, which is concerned with the history and development of ergonomics.

COMPONENTS OF ERGONOMICS

ANATOMY	Anthropometry	The dimensions of the body
	Biomechanics	The application of forces
PHYSIOLOGY	Work physiology	The expenditure of energy
	Environmental physiology	The effects of the physical environment
PSYCHOLOGY	Skill psychology	Information processing and decision making
	Occupational psychology	Training, effort and individual differences

WHO 91216

CHAPTER 1

The Provision of Energy

In order that a man can work or move about or even just stay alive in a relaxed position there must be energy available for his body to use. The body, like any engine, must operate within the limitations set by the laws of physics. The energy that is required in order that the muscles can exert forces must have previously entered the body in the form of food. In its simplest terms, the body is an engine that provides power from the locomotor system using oxygen that enters through the respiratory system to burn up the fuel entering through the digestive system.

Thus, when a man appreciably increases his rate of energy expenditure by working harder, the following effects inevitably occur. He will breathe more rapidly and more deeply in order to increase his intake of oxygen. His heart rate will increase as the automatic biochemical and nervous control systems of the body come into operation and it proves necessary to use the blood to transport more oxygen from the lungs to the muscles and move excess heat away from the working muscles to be dispersed around the body. This excess heat will also cause the body temperature to rise.

The ergonomist is committed to the use of a scientific and technical approach to the question of men working. He must therefore tackle the problem of human energy expenditure by looking for methods of measuring what is happening and describing the situation in terms of numbers. Of course, the problem is not so simple as it is in the case of mechanical devices. The great variety of muscles and joints within the body that can be used to perform work, results in large variations in the efficiency with which the body functions when regarded as an engine. This efficiency varies from about 10% to about 30%. Even at its best this may seem rather low, but in fact it is of about the same order as the efficiency of a steam engine. The range from 10% to 30% provides an opportunity but also creates a problem. It provides an opportunity because, if methods can be devised to ensure that a man uses his body at the upper end of the efficiency range rather than at the lower, he is able to do three times as much useful work for the same energy expenditure. It creates a problem because the amount of energy expended cannot be inferred directly from the amount of useful work done unless the efficiency is known.

An excellent example of the design of a machine that induces a man to adopt an efficient mode of energy expenditure is the bicycle. Admittedly this was not originally designed by an ergonomist—the design was arrived at by trial and error over many years—but it could in principle have been worked out by the following argument. The problem is to increase the efficiency of locomotion. Natural walking on level ground is highly inefficient. It is near the lower end of the 10-30% range quoted above. This is because energy is wasted in the reciprocating action of the legs, which are being accelerated and decelerated at every step, and in arm movements, which provide assistance in maintaining body balance. In addition, and probably most important of all, the head and trunk, which incorporate the main mass of the body, are moving up and down at each step to no useful purpose. Given this analysis the method of improvement follows logically. To keep the head and trunk at a constant height above the ground a seat and wheels are provided. This also reduces instability so that the arms need not move and can be placed on a control in the form of handlebars just to do the steering; the to-and-fro leg movements are replaced by the much more efficient circular movement using rotating pedals. Thus we have arrived at the bicycle and, incidentally, answered the question of why it is easier for a man to move along a road taking a mass of metal and rubber tubes with him than to walk unencumbered. It is because the bicycle is a machine beautifully designed to force the man into a very efficient mode of work. Thus in a given time he can cycle about three times as far as he can walk with the same energy expenditure.

The problem of measuring how much energy a man is generating can be approached by using the changes in state already described. That is to say, other factors being equal, the change in body temperature, in heart rate or in the rate of oxygen intake can be used as measures of physical effort. The main difficulty arises from the proviso that other factors must remain constant, since, in any normal working situation, there are many variables. For example, the bodily reaction to psychological stress, environmental stress or the stress of hard physical work is very similar. Thus, the fact that heart rate, respiration rate and body temperature increase may be an indication of the effects of hard work, heat stress or simply fear—or, more commonly, some unknown combination of all three.

The basic measure is oxygen consumption. This is normally calculated by measuring the volume of expired air and determining the remaining oxygen content by sampling techniques. Then, if air temperature and pressure are known, the volume of oxygen used in a given time can be ascertained, and from this the rate of energy expenditure, usually in calories per unit time, can be inferred. Even with the most advanced forms of instrumentation this technique requires considerable skill and experience on the part of the investigator and a high level of co-operation from the workers, who must perform their normal job· wearing face masks and carrying apparatus on their backs.

It is rather easier to measure body temperature or heart rate, but these measures are less direct and subject to greater disturbances by extraneous factors.

Even when a measure of energy expenditure has been obtained it has little meaning unless some standard of acceptable or tolerable effort is available. The work capacity of an individual varies with size, age, sex, nutrition, fitness, skill, and motivation. The standards are bound to have about the same range as body-weight. To arrive at an average, the long-term criterion is that the worker must not lose weight. Thus the standard is determined by food utilization or the available food intake. It is quite impossible for a worker to expend more energy than he receives in food for any long period, and working capacity in some countries is still capable of considerable increase by improvements in diet. Once this limitation is removed then it becomes a matter of assessing the maximum feasible rate of food utilization given that the climate itself does not impose a limit.

It must also be taken into account that there are two other possible limitations: maximum oxygen uptake capacity (i.e., aerobic capacity) and muscle volume or diameter. The body has elaborate mechanisms to compensate for oxygen debts due to sudden increases in activity, or temporary activity above aerobic capacity. These mechanisms are best reserved for athletics rather than for production work, except in emergencies. Even when the oxygen available is more than adequate for the required energy expenditure the work may be impossible because it must be done by muscles of insufficient volume to take up the required oxygen. Only the thigh muscles are large enough, as a single group, to utilize all the available oxygen input.

In summary, it is possible to measure physical effort and to express rate of working as a number, so that totally different jobs can be compared with each other and with average standards of acceptable work load. The value and range of application of this technique are self-evident. Nevertheless, obtaining these measures is skilled and tedious work, not to be undertaken except by trained investigators. The interpretation of results is even more sophisticated and requires the consideration of large numbers of variables that cannot be reduced to simple formulae.

General Principles

1. The body obeys the law of conservation of energy. All energy that appears as work must previously have entered the body as food. No worker can expend energy above the calorific content of his food intake without losing weight.

2. Even when the food supply is unlimited there is a limit to average energy expenditure set by the feasible rate of food intake. If this is exceeded the worker will lose weight.

3. Techniques are available for the measurement of physical work and the expression of the magnitude of energy expenditure in calories per

hour or per minute, which is quite independent of the kind of work performed.

4. Standards of acceptable levels of work based on the body's capacity to absorb oxygen can also be determined for particular populations. It is generally accepted that for performance over months and years with normal working hours the oxygen consumption at average work rate should be about one half to one-third of the maximum capacity.

5. Such standards may be above the maximum possible working rate for a particular job for a number of reasons—for instance because the physical load combines at times with excessive heat stress; because, despite an acceptable average rate, the load for a short time may be high enough to cause an excessive oxygen debt; or because the limit is set by the oxygen uptake capacity of the muscles doing the work.

6. Energy expenditure, heat stress, and psychological stress have effects on bodily mechanisms that are virtually indistinguishable by instrumentation techniques.

7. Only specialists should undertake the task of obtaining and interpreting measures of energy expenditure.

8. For normal work the heart rate should not rise by more than about 40 pulses per minute above the resting rate. The body temperature should not rise by more than 1°C above the resting temperature.

9. The maximum energy expended by a human operator is about 2 horsepower, but even in a fit, healthy young man this level of expenditure produces a large oxygen debt and can only be maintained for a matter of seconds. For day-long activity about one-tenth of this level—that is, 0.2 hp—is more appropriate. Given the low level of efficiency this means that useful work done by human muscles is inordinately expensive compared with mechanical power, even in countries where wage rates are low.

Particular Cases

1. A fairly high proportion of total available energy is needed for basal metabolism (that is, maintaining the body's internal functions) and for activity outside work, such as travelling to and from work. This means that, in particular circumstances, an increase in food intake may result in a proportionately larger increase in productivity. For example, suppose a job requires 300 calories per hour, if the average diet includes 3000 calories a day, about half of which is essential for purposes other than work, then the worker cannot do more than the equivalent of five hours' work per day. If a lunch of 750 calories is provided then the effective working day can be increased by 50% even though the calorie intake has been increased by only 25%.

2. In England the ratio of the cost of muscle-power to the cost of electric power is about 2000 to one.

3. Regular activity is essential for the maintenance of muscular strength. A completely unused muscle will lose half its maximum contractile power in a week.

4. Properly oxygenated blood is essential for brain activity as well as for muscle activity. If intensive physical work is stopped suddenly the blood supply to the brain may diminish and this may lead to loss of consciousness and collapse. Rhythmical muscular contractions help the heart to keep the required amount of blood in circulation.

THE PROVISION OF ENERGY

A. SOURCE OF MUSCULAR ENERGY

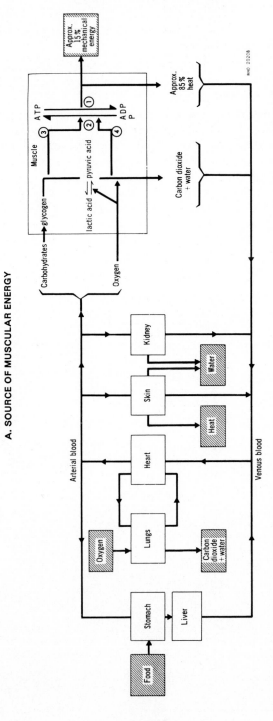

When oxygen debt occurs, reaction 4 is stopped, but reaction 3 continues until the concentration of lactic acid in the muscle cell inhibits reaction 2, which rebuilds the ATP necessary for the mechanical energy producing reaction 1.

B. FACTORS INFLUENCING CARDIOVASCULAR ACTIVITY

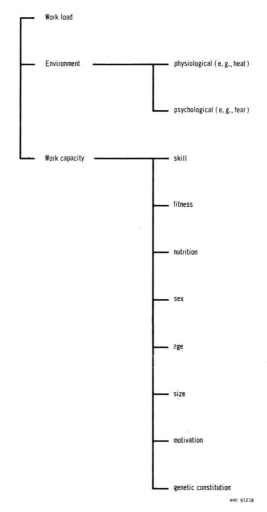

WHO 91218

N. B. The use of cardiovascular changes as a measure of work load is based on the assumption that all other factors are constant.

C. OXYGEN DEBT AND REPAYMENT

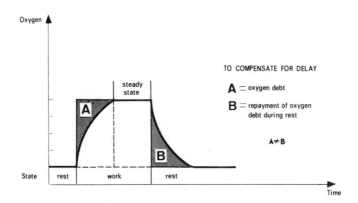

TO COMPENSATE FOR DELAY

A = oxygen debt

B = repayment of oxygen debt during rest

A ≠ B

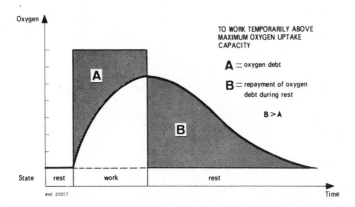

TO WORK TEMPORARILY ABOVE MAXIMUM OXYGEN UPTAKE CAPACITY

A = oxygen debt

B = repayment of oxygen debt during rest

B > A

WHO 20207

D. ENERGY EXPENDITURE ON PARTICULAR JOBS

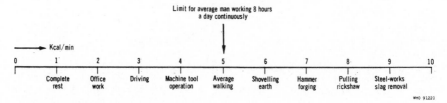

WHO 91220

1 kcal/min. ≈ 0.09 hp ≈ 70 watts

These are approximate values that vary with all the factors shown in Fig. B.

CHAPTER 2

The Application of Forces

Energy is turned into useful work through the locomotor system. For this purpose the body can be regarded as a set of rigid members: the parts of the limbs between joints, the main parts of the trunk, and the head. All these parts are quite heavy in relation to the available forces. Thus, unless care is taken, the available forces may be wasted in moving the parts of the body. Two consequences follow from this. Firstly, action will not be efficient if the direction of force is such that the limb or any other part of the body is moving against the force of gravity. Secondly, the most effective way of generating forces is to use the muscles and joints to get the body in a position such that the force is exerted by the body-weight rather than directly by the muscles.

For example, in lifting a weight off the ground, if the body and head are moving upwards against gravity as the weight is lifted, then most of the available force will be wasted within the body. Correspondingly, if a worker is required to exert a downward force in order, for instance, to fit an object into its surrounding part, then this task might well be accomplished more easily from a standing rather than from a sitting position. The weight of the head and trunk can be used in collaboration with stiff wrist, elbow and shoulder joints to provide the necessary force.

One other way in which body masses can be utilized is in the provision of forces from momentum. The force exerted is proportional to the mass of the moving part and also to the relative velocity of the limb and the object on which the force is to be exerted. The main use in working situations is to provide powerful forces for a very short time in sudden flexion or extension of the limbs. This principle is, of course, used extensively in boxing, where the art is in causing the opponent to move towards the fist at the time of impact, thus increasing the impact force by increasing the relative velocity, and also in getting the body-weight to move behind the fist, thus increasing the impact force by increasing the mass involved. There is always a danger that the forces generated might be greater than the body can tolerate. The most useful parts of the body for application of impact forces are the the soles of the feet, the sides of the hands, the shoulders and, for exertion against flat surfaces, the area across the shoulder-blades.

When a worker exerts some force as part of his job he is usually more concerned with precision and timing rather than with a maximum force. To achieve precision in limb movement he will normally use two sets of muscles about each joint—those concerned with causing the required movement (the agonists) and a complementary group opposing the movement (the antagonists). This may appear superficially to result in wasted energy but it is the best way to achieve high precision. Given that these two sets of muscles are required, then greatest precision will be achieved when the limbs are in the middle part of their total range of movement. It should also be remembered that human movement is inherently rotational and thus the motion which is simple and standard in engineering terms—the straight-line movement—requires the most complex integration of several joints on the part of a human operator.

There are invariably a large number of different combinations of joint movements which will effect the same result. It follows that an over-precise definition of the way in which a particular force should be applied is neither necessary nor even desirable, since a skilled operator will learn to vary his action so as to bring into use the greatest variety of muscle units and thus minimize fatigue effects. The same economy and spread of effort occur within a muscle. All muscles consist of large numbers of fibres and bundles of fibres that can be regarded as motor units. The greater the force or the faster the movement the greater will be the number of motor units involved. A slight change in the movement of a particular joint may result in a drastic change in the order, number and identity of motor units involved.

Another difference between engineering and anatomical problems arises in relation to static work, which for the human operator involves isometric muscle action. In this situation energy is being expended but there is no resultant movement, merely changes of the force exerted. This is unknown in mechanics, where the principle is used that no work is done unless a force moves its point of application. The human operator is certainly capable of work either by isometric muscle action or by isotonic action, in which there is limb movement. Each has its advantage and disadvantage. Isometric activity probably has superior kinaesthetic feedback—that is, the operator receives more information from the feel of what he is doing than from observing the results of the activity. Isotonic action, in which muscle fibres are moving in relation to each other, facilitates the blood flow and thereby the supply of oxygen and the removal of waste products. Thus for maximum precision isometric action is often best, but for maximum power output and for postponing fatigue isotonic action is usually superior.

Isometric activity is used extensively in the maintenance of body balance. This is often the critical factor in the application of forces. Without continuous muscle activity the body is not stable even in a sitting posture (one of the basic problems in the design of dummies is that if the flexibility of the human body is imitated such a dummy will slump grotesquely in any seat). When forces are exerted from a seated position it must be remembered that there will be a corresponding reaction on some part of the seat. For example, one important factor in determining forces exerted by the feet (e.g., pressing on pedals) is the design of the back rest of the seat. Even more difficult problems arise in relation to standing work, since the body is necessarily in unstable equilibrium and the maximum force is often restricted by the need to maintain balance rather than by joint and muscle limitations.

General Principles

1. For maximum energy availability use joints that involve large muscle masses moving over the largest possible distance.

2. For maximum economy of effort use the joints and muscles to produce a posture such that the body-weight exerts the required force.

3. For maximum resistance to fatigue from action:

    (*a*) use rhythmical movement;

    (*b*) provide for variation in joint action and combinations of joint actions.

4. For maximum precision use joints and movements such that effecting and opposing muscle groups are balanced.

5. For maximum instantaneous force use momentum—that is, the largest available part of the body mass moving with the highest available speed.

6. Large, steady forces depend on body stability rather than on maximum muscular contractions.

7. The principles of motion economy developed by work study practitioners are a useful summary of basic anatomical, physiological and mechanical principles in relation to work design. " Movements must be symmetrical " implies that there are no resultant forces disturbing the body balance. " Movements must be rhythmical " implies that energy shall not be wasted by excessive deceleration of limbs. " Movements must be natural " implies that only the muscle groups best fitted for the movement and natural joint positions are used.

Particular Cases

1. The basic principle of lifting is that the body-weight must be used to counterbalance the load and should, in the initial phase, move downwards as the load moves upwards.

2. From a seated posture it is never easy to exert downward forces.

3. The head is heavy, weighing 4-5 kg. Work in which it is not more or less balanced on top of the spine will be fatiguing. Minimum supporting muscular action occurs when the body is upright and the eyes are looking horizontally.

4. In order of increasing power and decreasing precision come the finger, wrist, elbow, and shoulder joints.

5. Pressure controls requiring isometric muscular activity can be superior to displacement controls for rapid, low-power, high-precision action.

6. An equivalent of the agonist/antagonist principle for a joint that gives high precision can be provided by using opposing limb actions as on a steering wheel.

7. Foot pressures for precision control should be exerted through ankle movements and not by the heel.

8. The maximum force from the human body is provided by double foot action from a seat with fitting back rest and hand grips.

9. The muscles involved in elbow flexion depend on the orientation of the hand; the greatest force is available when the palm is facing the shoulder.

THE APPLICATION OF FORCES

A. ASPECTS OF WEIGHT DISTRIBUTION

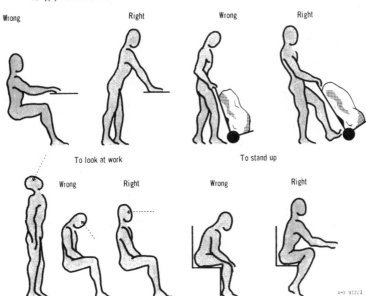

To apply a downward force

Wrong   Right

To lift on a trolley

Wrong   Right

To look at work

Wrong   Right

To stand up

Wrong   Right

B. LIFTING AND CARRYING

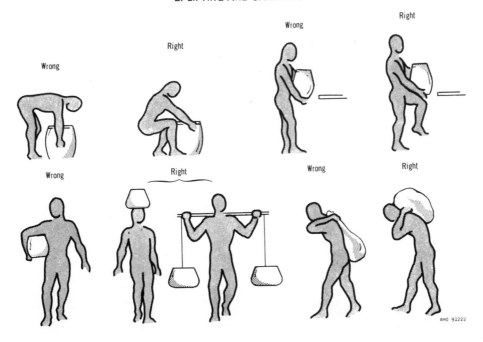

Wrong   Right

Wrong   Right

Wrong   Right

Wrong   Right

C. USE OF HAND CONTROLS

For maximum force

with some precision

controls allowing
double arm action
with use of shoulder
and upper arm muscles

with negligible precision

controls allowing
use of back shoulder
and upper arm muscles

For maximum precision

with some force

controls allowing
double arm action
with use of upper
arm muscles

with negligible force

lower arm fixed
support with wrist
and finger action

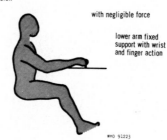

WHO 91223

N.B. Stable footrest required

D. USE OF PEDALS

For maximum force

well-designed
back rest

hand-grips

160°

For maximum precision

in unstable conditions

in stable conditions

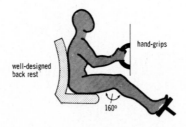

rudder bar operated
by ankle movements

treadle operated
by ankle movement

or by double foot action

WHO 91224

Notice the additional force applied to the tree through the shoulder as well as through the lever.
Reproduced by permission of the Forestry Commission, Farnham, Surrey, England.

CHAPTER 3

Problems of Body Size and Posture

Although the designer of tasks can often rely on the flexibility of body posture and the adaptability of human behaviour, he must accept the limitations of fixed size. Thus the starting point of the design of work spaces must be the dimensions of the people who are going to operate within those spaces. One of the primary responsibilities of ergonomics is to provide data about body size. This study, which is part of the province of the anatomist, is called anthropometry.

As a problem merely of mensuration it may appear, at first sight, to be quite straightforward and capable of solution by relatively little research effort. Unfortunately this is not true, and such a study turns out to be as complex and tedious as any other biological investigation. There are many complicating factors.

To begin with there are no easily definable points consistent with any high degree of accuracy. To take the most obvious one of height, the length of a man is not necessarily the same when he is standing up as when he is lying down, since in the standing position he can alter his height appreciably by, for instance, changes in muscle tension and spinal curvature. These variations are not large enough usually to be of any consequence in work design but they are tiresome from the point of view of scientific precision and simplicity. More serious is the question of precise definitions in relation to limb sizes. For example, if we wish to describe arm length where are the end points? Do we start from the flat of the back, or from the axis of rotation of the shoulder or from the front of the chest? Do we end at the wrist joint, the grasping point, or the finger extension point? There are no standards or even generally accepted practices among anthropometrists, so each set of useful data must include detailed statements of measuring techniques and definitions of end points.

The variations between individuals are usually large enough to be important and so the statement of averages is not enough. Ranges are required, but since human dimensions are distributed normally in a statistical sense there are very few individuals at either extreme, most individuals being much nearer the average. The question therefore arises of what proportion of actual ranges of size should be taken into account in design. For example, most adults are between 1.50 m and 1.85 m tall but some men exist whose height is as much as 2.50 m. It would clearly be absurd to make all doorways at least 2.50 m high to cater for a few individuals in the total world population.

In this context it is now standard practice to take into account 90% of the population, omitting the upper and lower extremes of the dimension under consideration. For this reason anthropometric data are now often expressed in the form of 5th, 50th and 95th percentiles. A dimension quoted at 5th-percentile level means that 5% of the population considered are estimated to be smaller than this. For a normal population the 50th percentile is the same as the average: 50% of the population are below and 50% are above the size quoted. Correspondingly only 5% of the population are larger than the 95th percentile. The range from 5th to 95th percentile covers 90% of the population.

It is necessary to quote many dimensions separately, since it has never been accepted that other body dimensions can be expressed as a given percentage of height. Individuals are obviously not entirely consistent: there are tall men with short legs and short men with long arms, but it can be argued that these are sufficiently rare to be ignored for many design purposes and that dimensions as proportions of height could be used effectively if the data were available.

This would be a useful simplification, since it is not possible to use one table of anthropometric data for all work-space designs. Again to take an extreme example, it would not do to use anthropometric data obtained from young soldiers for the design of a work space to be used in a factory by middle-aged female workers. One of the practical problems of anthropometry is that there is a surfeit of data about fit young servicemen and not enough about other sections of the population.

Apart from sex differences there are also important anthropometric variations between ethnic groups, geographical areas, social classes and even occupations. Yet another source of variation can be traced to improvement in diet. There have been significant increases in body size among working populations in Europe since the Second World War, and corresponding changes are bound to occur in developing countries as dietary levels and balances change. It is reasonable to suggest that anthropometric data for such populations will need to be checked about every 10 years or so.

It will be clear that even when anthropometric surveys have been conducted and the raw data are available there are still problems of presentation and application. Some of these stem from the fact that the compilers of the data usually have a background in medicine or biology, but the users have a practical industrial or engineering background. It is of little use to the work-place designer to know that the popliteal height is 40 cm, unless this information is presented in such a way that he can appreciate its relevance to the height of seats. This problem of specialist terminology can be overcome by the proper use of diagrams or manikins. For this purpose a manikin is a scale model of a human operator, usually in two dimensions only, with swivel points corresponding to the sites at which joints permit movement. These can be made for different percentile levels so that the question of ranges is nicely dealt with.

There are two problems that are not covered either by the compilation of tables of data or by the use of manikins. These are to do with tolerance and with dynamics. As regards tolerance, there are always dimensions in a particular work space that are not critical to within a few centimetres and others where a centimetre might be important. How to distinguish matters of priority in fixing dimensions is something that the anatomist has found difficult to communicate to non-specialists. Similarly with dynamic problems, the operator is not fitted into a work space as a static entity—he must move and manipulate. The extent and frequency of these actions can affect the optimum work-space dimensions.

Two methods of utilizing anthropometry that partially take these factors into account have been developed. The first is to design and present data for a special problem. For instance, a chart is available that indicates the area of acceptable centre-line heights for lathes as a function of certain design variables. The task of translating the raw data provided by anthropometrists into a form directly usable by lathe designers has been carried out by an ergonomist, who took into account such factors as variations in size of likely user populations, the need to manipulate controls and materials, and so on. In this case the range of application of the data has been reduced in order to make them more readily communicable.

A more general technique is to use what are called " fitting trials ". This is effectively the use of live subjects as manikins. In its simplest form a full-scale model of the work space is built and various people from the user population try it, their behaviour and comments being taken into account in subsequent modifications of the design. In its most sophisticated form this technique requires the proper sampling of subjects from the user population and elaborate design of their tasks and of the procedures for acquiring dimensional data from them, but the power of the method lies in its versatility. The design of the study can be as simple or as complicated as required, according to the degree of accuracy considered necessary in determining work-space dimensions.

In parallel with the fixing of the dimensions of the work space the designer must make basic decisions about operator posture. For example, should the work be done from a seated or a standing position? Many factors come into this decision. The range of working area in any plane is greater for the standing than for the sitting operator, and if frequently used controls must be scattered over a large area then the operator must stand. There is no point in providing a seat at a machine (e.g., a console or a desk) unless the operator has adequate knee room under the working top; without knee space he is likely to be more comfortable standing. As mentioned in the previous chapter the application of forces is much easier from a standing position. Yet another factor in favour of standing is that the worker, if he has to move away from the main work space very frequently, may well use more energy in getting in and out of his seat than he would do in standing while working. Many of these problems can be overcome by using moving seats or just high seats with footrests.

Since the design of seats is a universal problem that has been widely studied, the principles are reasonably well, if not entirely, understood. They depend on the maintenance of an equilibrium that allows not only the appropriate range of overt actions but also the proper functioning of such body processes as seeing, respiration, circulation and digestion. The provision of support in areas that need it and can tolerate it requires anatomical and physiological knowledge.

This is not to say that one needs to be an anatomist and a physiologist to be able to consider postural aspects of work spaces, and conversely a knowledge of classical anatomy and physiology is not enough. The skill required involves having not only a general appreciation of the background biology but also a global sensitivity to the mechanics of limbs and structures and the ways in which work spaces can provide the freedom, the flexibility, and sometimes the incentive for the proper use of the human body.

General Principles

1. The dimensions of work spaces should be matched to 90% of the possible user population.

2. In selecting anthropometric data the designer should take particular note of the exact definitions of the measurements (especially the end points), the size and kind of population sample from which the data were obtained, and the availability of range information in addition to average information.

3. In using anthropometric data the designer should take particular note of the acceptable tolerance with regard to the different dimensions and the difference between static fit and dynamic fit.

4. Optimum matching of dimensions may alter with time as population size changes and as the kinds of workers doing the particular task change.

5. Broadly there are four vehicles for the use of anthropometric data:
   (a) tables and drawings;
   (b) manikins;
   (c) graphs and nomograms for particular tasks; and
   (d) fitting trials.

6. Postural problems are inextricably bound up with size problems so that the two must be considered together.

7. Decisions on seating *versus* standing must take account of:

(*a*) the location of controls, components and activities;

(*b*) the available knee room;

(*c*) the size and directions of forces to be exerted; and

(*d*) the frequency of standing and sitting.

8. Design of seats must take particular account of:

(*a*) relationship between seat and work area;

(*b*) variability of posture;

(*c*) ease of standing and sitting;

(*d*) stability, particularly if the seat is movable; and

(*e*) optimum cushioning of seat and back rest.

9. Just as a comfortable seat provides for variation of posture so also does a well-designed standing work space provide for ease of movement. Restriction of movement usually increases fatigue. It is often possible to design a work space so that the operative has a choice of sitting or standing.

Particular Cases

1. The 5th and 95th percentiles rather than the 50th (the average) may determine work-space dimensions; for example, the maximum height of controls depends on 5th-percentile upward reach, and the minimum height on 95th-percentile knuckle height.

2. Economic class is one important determinant of size. For example, in Great Britain the average height of male company directors is about 1.80 m while that of male unskilled workers is about 1.70 m.

3. The dimensions of a work space—e.g., a tractor seat and controls—properly calculated for workers in a particular country may be quite unsuitable for workers elsewhere, since human dimensions vary considerably from one country to another.

4. A well-designed seat does not necessarily result in the most relaxed posture. The criticism that well-designed seats send workers to sleep is not valid.

5. The reduction of postural problems by bringing controls closer together may increase problems of identification.

6. Postural problems may arise in relation to visual information as well as in bodily actions—as, for example, in the case of the posture assumed by a truck driver in backing his loaded vehicle.

7. There is often a conflict between the optimum posture for manipulation and that for vision—e.g., in fine manipulative work the best position for the hands is not the best for seeing and *vice versa*. The usual but not universal solution is to put the work near the eyes and provide supports for the arms.

BODY SIZE AND POSTURE

A. MAIN HUMAN DIMENSIONS TO BE ASCERTAINED FOR DESIGN OF WORK SPACES

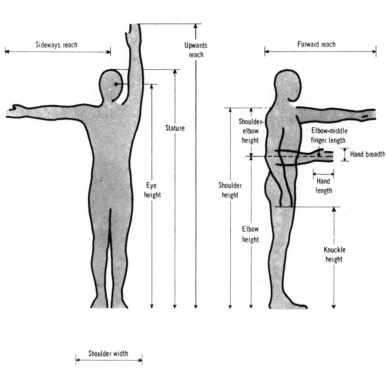

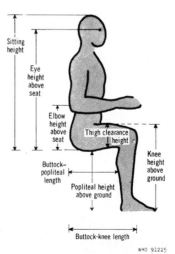

WHO 91225

B. IMPORTANCE OF RANGE OF BODY SIZE IN DESIGN

Maximum control height upward reach of short man	**Minimum control height** knuckle height of tall man	**Clear visibility** eye height of short-bodied man	**Roof height** height above seat of tall man
Seat height height under knees of short-legged man	**Seat width** buttock width of fat man	**Door height** suitable for tall man	**Maximum reach** suitable for small man

WHO 91226

C. POSTURAL ASPECTS OF SEATING

BODY PART		DESIRABLE POSTURE	DESIGN IMPLICATIONS
ALL		Small variations of all postures should be possible including movement of inactive parts	All aspects of work space should allow for these minor changes
HEAD AND NECK		Head should be more or less balanced on shoulders with horizontal line of sight. Extensive, frequent or rapid head movements should be avoided	Main display elements should be at eye height or below. Angular separation of display elements should be limited
TRUNK		More or less vertical with normal spinal curvature to minimize effort of stabilizing musculature, facilitate breathing and maintain maximum stability	Provision of back rest and appropriate location of display and controls
UPPER LIMB		Upper arm roughly vertical. Forearm roughly horizontal. Wrist such that palm of hand faces downwards and inwards	Provision of arm rest and placement of hand controls
LOWER LIMB		Thigh roughly horizontal. Lower leg at obtuse angle to thigh. Foot at right angles to lower leg	Size, height and slope of seat; provision for adjustments and perhaps foot rest. Placement of foot controls

WHO 91227

D. SEATING AT THE WORK SPACE

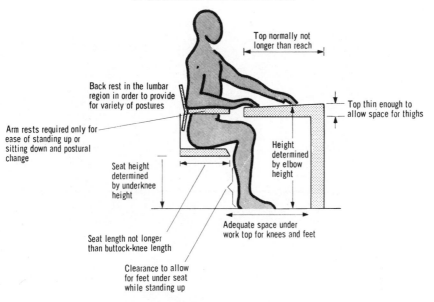

Top normally not longer than reach

Back rest in the lumbar region in order to provide for variety of postures

Top thin enough to allow space for thighs

Arm rests required only for ease of standing up or sitting down and postural change

Height determined by elbow height

Seat height determined by underknee height

Seat length not longer than buttock-knee length

Adequate space under work top for knees and feet

Clearance to allow for feet under seat while standing up

Seat surface slightly tilted backwards and not smooth to maintain equilibrium

Seat width determined by sitting buttock width

WHO 91228

E. WORK SPACE OF A TRACTOR DRIVER

Reproduced by permission of Massey-Ferguson, Coventry, England.

F. COMPLEX POSTURE REQUIRED FOR TRACTOR OPERATION

The driver has to operate simultaneously hand and foot controls on one side of the seat and hand controls behind the seat while also receiving visual information from behind.

G. CUTTING BRANCHES WITH A POWER SAW

Complex postures are required because of uneven ground, variety of branch positions, and saw weight.
Reproduced by permission of Hyett Adams Ltd., Stonehouse, Gloucestershire, England.

Note the unusual development of back muscles resulting from the permanent bending posture.
Reproduced by permission of the Office of the High Commissioner for Australia, London, England.

CHAPTER 4

The Effects of Climate

Change of body-core temperature in the human being can no longer be tolerated when the rise or fall exceeds a small percentage of the natural ambient temperature variations between different parts of the world, seasons of the year, or even times of day in a particular place. Conditions of work—e.g., with molten metal or with refrigeration—can generate additional stresses, so that the consideration of climatic problems is an essential part of ergonomics. As always, we must define the problem accurately and study it in numerical terms.

In physical terms the situation is straightforward: the body obeys the elementary laws of physics of heat exchange. In this terminology a man can be regarded as a black body—that is, the body surface, quite independently of the skin colour, is a very good absorber of radiant heat and also a good radiator of heat. He is usually either gaining or losing heat by convection and radiation depending on whether the surroundings are colder or warmer than his skin. Heat is also gained by metabolism and lost by sweat evaporation.

In biological terms there are a variety of mechanisms by which the body temperature is maintained. As a first line of defence only the core temperature of the body needs to be kept within narrow limits. In most areas, the body can tolerate considerable temperature variations, to a depth of about 2 cm below the surface, and this in fact constitutes about half the total body mass. The skin temperature can vary by as much as 15°C above or below the core temperature for a short period without irreversible harmful effect. The second line of defence is provided by the nervous and endocrine automatic control mechanisms of the body, which come into operation when the core or skin temperature changes. The main effects are increases in circulation and sweating, in the case of exposure to heat, and initial reductions and later increases in circulation in the skin (gooseflesh), increased metabolism and shivering, in the case of exposure to cold. The third line of defence is in behavioural and postural changes, which can have a considerable influence in modifying exposure to radiation and the effects of convection—but a worker is often not free to use these methods since his posture is determined by his task. The fourth line of defence involves an intelligent approach to the question of exposure in the design of appropriate clothing and the creation of a protective environment, ranging from primitive means of providing fire and shade to the installation of air conditioning.

Because of the activities of the body's automatic control mechanisms the translation of physical measures of the environment into measures of heat stress is not easy. To take a few simple examples: a warm, humid environment can feel more unpleasant and perhaps interfere more with work than an even warmer but dry environment; a moderate air temperature will not accurately indicate the extent of heat stress if there is a large and very intense source of heat radiation, such as a furnace, near the operator; a very hot day with a breeze blowing is preferable to a hot day with still air; by contrast a cold windy day is much more unpleasant than a very cold calm day.

In more scientific terms we can only obtain a measure of climatic stress by measuring all the physical variables that affect the rate of heat exchange. Convection rates depend on air temperature and wind speed. Radiation rates depend on the surface temperature of the surrounding objects and walls. Evaporation efficiency depends on air temperature, wind speed and humidity. The metabolic rate depends on how hard the work is. Thus the following measures are needed:

(a) air temperature measured by a shaded thermometer;

(b) radiant temperature measured by a globe thermometer, which is an ordinary thermometer with the bulb at the centre of a six-inch black copper sphere;

(c) humidity indicated by the difference between dry-bulb and wet-bulb temperatures and expressed in terms of relative humidity or preferably as vapour pressure; and

(d) wind speed measured by a kata thermometer (which works on the principle that the rate of cooling of a thermometer bulb is a function of wind speed at a given air temperature), or by a hot-wire anemometer.

Many attempts have been made to combine these measures into a single index of heat stress. The simpler ones, such as *effective temperature*, correct for the effects of different humidities and wind speeds by estimating (on the basis of opinions expressed by subjects exposed to different environments) the temperature of still saturated air that is felt to be equivalent to the measured environment. The additional effects of radiation are incorporated in the *corrected effective temperature* scale. The allowances made for different clothing in these scales are not very satisfactory, and no account is taken of rates of working, which can increase the amount of metabolic heat generated by a factor of 10 or more. In naval service it is common to have to work for 4 hours under heat stress conditions, and for this reason the *predicted 4-hour sweat rate* (P4SR) was developed during the Second World War. Again, this is based on experimental studies, in this case a calculation of the amount of sweat lost under particular conditions. Clothing and working rate were included as variables in these studies and are incorporated in the resultant nomograms. Two more recent measures attempt to provide a *heat stress index*, varying from 0 to 100%, by calculating the sweat evaporation required to maintain thermal equilibrium under the measured conditions as a proportion of the maximum possible evaporation rate. It will be noted that all these measures are based on the assumption that the sweat rate is proportional to the heat load due to the environment and to the physical work performed.

The advantage of these measures is that they can be used to assess working conditions in terms of whether these conditions are tolerable for continuous work or not and, if not, for how long the worker could be expected to remain in such conditions and how long it will take him to recover when he emerges. In practice it is still necessary to check these estimates from direct measures of core temperature, heart rate, and sweat loss on men actually working in the environment under consideration. There are quite sophisticated instruments available to perform these measurements continuously during the whole working day without interfering with the workers' activities. Such instruments utilize electronic and telemetric techniques. However, in everyday practice core temperature can be measured by a clinical thermometer, heart rate by counting the radial or precardial pulse rate during short rest pauses, and sweat loss by the change in body-weight during the working day, taking into consideration the gain due to fluid and food intake as well as the loss due to urinary and faecal evacuation. Similarly, if the problem is to change a climatic environment, it is more important to know the actual environmental parameters, such as air temperature, radiant heat, air velocity and humidity, than to obtain some combined measure.

Little is known about the effects of continuous heat stress for longer periods of time. It is known, of course, that steps must be taken to replace water and salt lost in sweat, but there appear to be large individual differences in human tolerance. In particular the conditions and duration of heat exposure that result in the breakdown of the heat regulatory mechanisms, giving rise to heat exhaustion, cramp or stroke, vary from one individual to the other. Most of the studies have been done on fit young men and little is known about the effects of age, sex, fitness, and habituation, except in the broadest terms. Similarly, differences of race, body build, personality, and so on, have not been explored systematically to the point where any firm conclusion can be drawn. It is known that the colour of the skin makes some slight differences to the thermal properties, but there appear to be no significant differences in heat tolerance between Negroes and Caucasians. After approximately one week of continuous work in heat, man becomes acclimatized to the conditions if they are within the limits of his tolerance. As a consequence he will feel less discomfort and his performance may improve. The mechanisms underlying acclimatization are again not fully understood, although lower salt concentrations in the sweat, increased sweat rates at lower skin temperatures, and reduced heart rate and body temperature increases have been observed after acclimatization.

It is useful to make the distinction between stress and strain in a manner analogous to the way in which the terms are used in engineering. Stress is the load imposed by the climatic conditions and physical work, and strain is the resulting effect. There are two basic difficulties; one is that there are many dimensions to the concept of strain, and the other is that stresses are sometimes used as measures of strains and *vice versa*. There are at least three kinds of strain each with an array of possible measures. The first is the subjective experience of the climatic variable: it may be comfortable/uncomfortable, tolerable/intolerable, depressing/exhilarating, and so on; it may also be compared directly with another set of stress parameters—e.g., a man can be asked which temperature of saturated still air is " the same " as particular temperature/humidity/wind speed combinations. The second is to measure changes in the performance of particular tasks under particular changes of stress—e.g., one could study the effect of humidity on the ability to do mental arithmetic or that of air temperature on creativity. The third is to measure changes in bodily functions that cannot be controlled at will. Some of these, such as skin temperature, core temperature and heart rate, are obviously responding directly to heat stress, but for others, such as alpha rhythm and heart regularity, the relationship is indirect. There seems to be some evidence that mild heat stress is arousing but excessive heat stress, like cold stress, reduces the normal level of arousal.

Given all these possible permutations and combinations of studies and measures it is not surprising that progress is slow. The present position seems to be that opinions on optimum or comfortable climatic conditions vary widely with race, sex, age, acclimatization, cultural background, and so on. Females rather than males and older rather than younger people prefer higher temperatures for comfort. Studies of performance, concerned with, for example, speed, accuracy, and accident occurrence, confirm that there is deterioration at high and low temperatures, but effects are slight until climates become extreme. In other words, for healthy people acceptable climatic conditions form a fairly broad plateau rather than a peak.

Optimum environmental conditions depend mainly on the rate of physical work; obviously for a higher rate of heat generation a lower ambient temperature will be desirable. In normal working conditions such as office work, where energy expenditure is not high, a temperature in the region of 19°-23°C, a humidity of 30-70%, and an air movement of 3-12 m/min will be acceptable to most people accustomed to a temperate climate.

In essence the core temperature of the body must not increase above 38°C in prolonged daily exposure to heat and intense physical work, the pulse rate should not rise above about 110, and the diet—particularly intake of fluids and salt—must be adequate to replace losses. The thirst mechanism is not very sensitive, so that people exposed to heat must be encouraged to drink more. Weight loss due to dehydration should not exceed 2-3 % of body-weight.

The problem of cold is easier to deal with, both conceptually and practically. Small falls in body temperature can be tolerated for a time with much less strain than can rises. Nevertheless falls in core temperature must not be accepted. Under conditions of work they can be prevented by proper clothing. The most difficult task is to protect the toes, fingers, ears, and nose from chills. Cold fingers lose their sense of touch and motility and become numb; as a consequence the worker's skill is diminished and the danger of accidents is increased. The bulkiness of too much clothing limits its utility; under extremely cold working conditions work has to be interrupted for respite in a heated area. Exposure to cold conditions results in a considerable increase in food requirements to allow for increased metabolic rate and, over the long term, to provide an increased subcutaneous fat deposition, which is a very good heat insulator. There is a cold stress index analogous to heat stress indices called the *wind chill scale*. As the name implies this is very much a function of wind speed at a given air temperature. It is measured by the time taken to freeze water in a metal container. There is also a unit of clothing insulation called the clo, which is defined in physical terms (a heat loss of 5.5 kcal/m²/hour across a gradient of 1°C, or 0.18°C/kcal/m²/hour), but the scale is fixed roughly so that 1 clo represents the average insulation of a comfortably clothed person in a normal environment. The principles of clothing design for insulation are well understood. The basic principle is to provide a layer of still dry air on the skin and to increase the thickness of this layer as improved insulation is needed. The air must be kept still and dry by an outer layer that is wind-proof and rainproof (but if possible water-permeable to disperse sweat). The characteristics chiefly required in the internal material, which prevents lateral air movement and maintains the air gap, are that it must be of low density and must not retain sweat.

General Principles

1. Any measure of heat stress must take account of air temperature, radiant temperature, air velocity, humidity, and the intensity of work to be performed.

2. There is no entirely satisfactory single measure of heat stress but various measures are available for different ambient and working conditions.

3. Tolerable heat conditions can be assessed by using the man himself as a sensing device. Sources of information are:

(*a*) his opinions;

(*b*) his thermal state;

(*c*) his physiological responses; and

(*d*) his performance.

4. Heat stress, which is the load on the man, should be distinguished from heat strain, which is the effects of the load.

5. The optimum environmental conditions for different activities cannot be defined with great precision, since there are differences within an individual at different times and between individuals at the same time.

6. The skin temperature can vary widely but the core temperature of the body should be kept fairly constant.

7. The main factor, other than temperature, that produces coldness is wind speed. Thus, to reduce cold stress, the worker should be protected from the wind.

8. The effectiveness of clothing as a protection from cold depends on heat insulation, which is markedly reduced by dampness.

9. The basic principles of clothing design for hot conditions are to increase the body area from which sweat can be evaporated and to provide shading against radiant heat.

10. Climatic conditions cannot be considered in isolation from other forms of psychological and environmental stress.

Particular Cases

1. Mild heat stress can actually improve performance at mental tasks for a limited time, but high heat stress or cold stress will have the reverse effect.

2. Workers usually automatically take up the most advantageous posture. Changes in posture can markedly affect convection, radiation and evaporation.

3. There is no evidence that radiation from the sun can have peculiar effects on the head or spine apart from its effect as an intense radiant heat source. The ultraviolet radiation is absorbed in the epidermal layer of the skin.

4. Reduced temperature of the extremities can change performance although core temperature remains unchanged—e.g., when the hands are carrying out complex manipulations.

5. A fit man can lose more than 2 litres of sweat per hour for a limited time. A maximum sweat loss of 10-12 litres during 8-hour work shifts may be tolerated by acclimatized workers over periods as long as several months yearly (summer season).

6. A reduction in climatic stress may not be welcomed by some workers, since it may lessen the justification for additional bonuses.

EFFECTS OF CLIMATE

A. EFFECTS OF HEAT STRESS

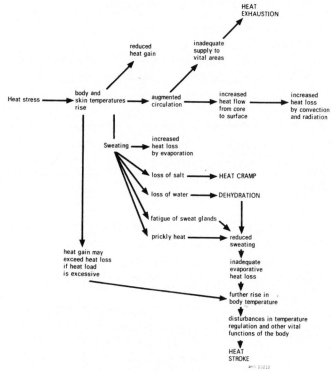

Adapted from an original by H. S. Belding.

B. CHANNELS OF HEAT LOSS

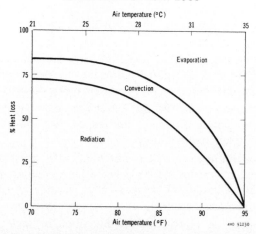

Under conditions of still air and low humidity, and with a mean radiant temperature of 35°C.
Adapted from an original in: *Life sciences data handbook*, published by the United States National
Aeronautics and Space Administration, Washington, D. C.

C. THERMAL LIMITS

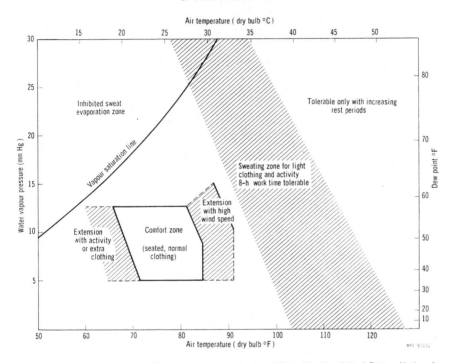

Adapted from an original in: *Life sciences data handbook*, published by the United States National Aeronautics and Space Administration, Washington, D. C.
N. B. These data are approximate and subjective. They were obtained mainly from North American subjects of European origin, but there is no evidence that race or colour makes any appreciable difference.

D. HEAT STRESS INDICES

SCALE	BASIS	OPERATOR CONDITIONS	ADVANTAGES AND LIMITATIONS
Corrected effective temperature (CET; Yaglou, Bedford)	Stress equivalent to still saturated environment at this temperature. Nomograms available	Full indoor clothing (normal scale) or stripped to waist (basic scale). No allowance for varying work rate	Permits expression of air temperature, humidity, radiant heat and air velocity in one figure. Does not include metabolic heat. Different climates sharing the same CET are not necessarily equivalent from the point of view of physiological strain, particularly below 40% relative humidity
Predicted 4-hour sweat rate (P4SR; McArdle)	Average sweat loss of fit acclimatized young Europeans in this environment for 4 hours. Nomograms available	Variation of clothing and rate of working included as variables in nomogram. Continuous heat exposure	Permits estimation of anticipated sweat loss and consequent water replacement requirement. Inaccurate below 40% relative humidity. Not applicable to variable intermittent heat exposure. Validated only on young acclimatized men
Heat stress index (HSI; Belding and Hatch)	The evaporation required to maintain thermal equilibrium expressed as ratio of maximum possible evaporation. Equations with heat exchange coefficients and nomograms available	Originally for semi-nude men with corrections for light clothing; for various work rates and continuous heat exposure	Permits estimation of required evaporation and consequent water replacement requirement. Permits estimation of maximum possible evaporation of sweat. Permits estimation of tolerance time and required resting time. Difficult to apply to variable, intermittent heat exposure. Validated only on young acclimatized men
Index of thermal stress (ITS; Givoni)	The sweating required to maintain thermal equilibrium. Equations with heat exchange coefficients and nomograms available	For all combinations of work rates and clothing	Estimation of required sweat loss and consequent water replacement requirement. Otherwise the same advantages and limitations as HSI.

WHO 92232

Relevant physical measures: air temperature (dry bulb), mean radiant temperature (black globe), and air velocity (kata thermometer or anemometer).

E. HEAT STRESS AND STRAIN

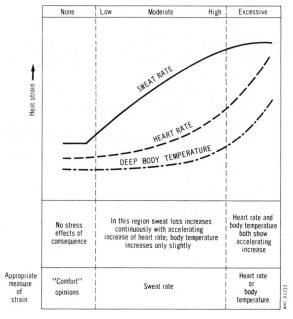

F. HEAT AND WORK STRESS

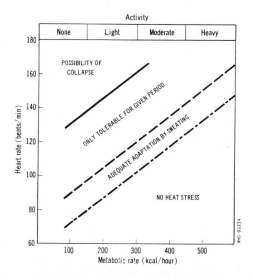

G. EFFECTS OF COLD STRESS

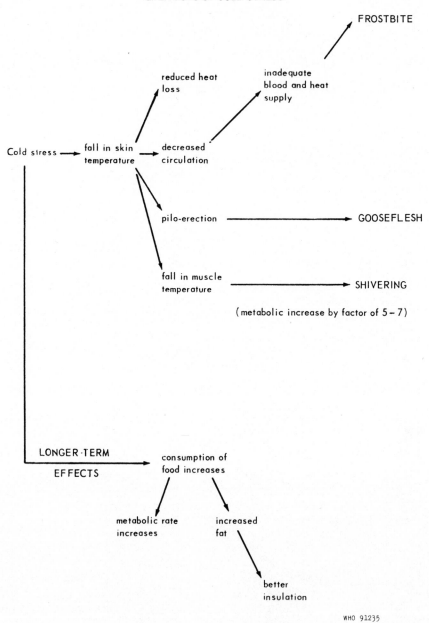

(metabolic increase by factor of 5 – 7)

WHO 91235

Adapted from an original by H. S. Belding.

H. PROTECTION FROM COLD

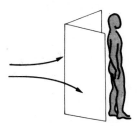

Erect artificial
wind barrier

Ensure that extremities
including head are well
insulated

Keep dry
by tent

Use natural wind
barrier

Ensure that wind cannot
penetrate clothing at
natural gaps (neck, waist,
wrists, ankles)

Keep dry by
plastic raincoat

WHO 91236

CHAPTER 5

Limitations of the Sense Organs

One condition for effective work is that the worker must control his activities on the basis of information he receives from his surroundings. Although the intact healthy man has a highly versatile array of sensory mechanisms, each of these mechanisms has limitations. Within the normal range of operation there are many conditions and variables that affect efficiency. Above a certain limit the channel of information becomes a source of strain and eventually a hazard in that the mechanism is susceptible to damage. Thus, for each sensory mechanism there are two related problems of environmental design: to determine under what conditions the mechanism operates best as an information channel and under what conditions it becomes a source of strain.

There are some senses, such as smell and taste, of which the usefulness in work is relatively rare. In broad terms there are three senses of importance in work design: the visual system, the auditory system and the kinaesthetic system. The corresponding environmental variables are light, sound, and vibration and motion. Sound is, of course, a special kind of vibration—special because there is a specialized sensory channel. From the operator's point of view the detection of vibration depends, in the first instance, on whether or not the body moves. Body motion is detected by the *vestibular system* located behind the ear. Angular acceleration is detected by the semicircular canals, of which there are three, arranged in three planes at right-angles. Movement of a fluid causes the bending of hairs, which is detected by associated nerve endings. Thus angular acceleration in any plane will be perceived through at least one of these organs. Linear acceleration is detected by the utricle and saccule, which are roughly spherical in shape and operate in a similar way to the semicircular canals, except that the hairs contact calcium carbonate blobs, and it is the movement of these small masses that is detected as the response to linear acceleration. Since these mechanisms are part of the proprioceptive/cerebellar system the operator is not directly aware of stimulation. Internal parts of the body—the intestines, for example—also move in response to acceleration and vibration, but there is no evidence that such movement causes any problems until it is sufficiently severe to result in damage. Motion sickness is probably due to activity of the vestibular system in the range of 0–30 cycles per minute. Above 60 cycles per minute—that is, in the range of 1–500 cycles per second—movement is perceived as vibration rather than motion.

If the body is not in motion then detection of lower frequencies depends on the *touch system*. Up to about 20 cycles per second these receptors provide a sensation of discrete pressures in rapid succession, but at about that frequency this type of sensation is gradually transformed into a sensation of vibration. This vibration range is 20–250 cycles per second. The use of vibration deliberately provided to convey information— e.g., for aircraft pilots—has been tried but without much success in practice. This is not to deny that natural vibration is an important source of information, as it clearly is, when used by experienced operators in the control of all forms of transport and in the operation of machine tools.

The *auditory system* functions in the range 20–20 000 cycles per second, although the upper limit may decrease to 12 000 cycles per second with increasing age. The anatomy of this system is complex, since it involves the conversion of pressure changes in air, through pressure changes in a liquid medium, into pulses, which contain pitch and loudness information. In broad terms a cylinder (the external auditory meatus) protects the ear drum and maintains it at a constant temperature and humidity. Movement of this drum (the tympanic membrane) is transmitted through a series of mechanical linkages (the ossicles) to another drum (the oval window) and then through a fluid in the cochlea to a resonator (the basilar membrane) and its associated sense cells. Different parts of this membrane respond differently to different frequencies and amplitudes. Study of the associated phenomena is unduly complicated by some confusion of terminology. The changes in the environment are in frequency and amplitude, while the changes in sensation are in pitch and loudness: there is some correspondence but not an exact one. The term " sound " is used indiscriminately to describe the physical variables and the sensations. Thus measurement of sound might be legitimately interpreted as a problem for a physicist or for a psychologist depending on which meaning of the word is intended. The most commonly used unit, the *decibel*, is a physical measure of sound intensity on a logarithmic pressure ratio scale. There is confusion here because the scale and its zero point are chosen to correspond to subjective phenomena, although it still remains a physical scale based on amplitude and is similar but not identical to a measurement of loudness. Loudness level is measured in phons. The *phon scale* takes account of variations in loudness with frequency as well as with amplitude and is identical to the decibel scale only for tones of 1000 cycles per second. It is not linear in subjective terms (40 phons is not twice as loud as 20 phons). The *sone scale* provides a unit of loudness that does meet this criterion. The loudness level corresponding to 1 sone is 40 decibels or phons at 1000 cycles per second. For each phon level there is a corresponding sone level; the phon range of 40–150 has a corresponding sone range of 1–2000. Measures of sound intensity limits stated in terms of decibels must be interpreted with the above provisos in mind.

The limit of acceptable loudness can be plotted on a decibel/frequency graph. This represents the points at which permanent damage may occur. Such damage is not usually revealed by increased deafness over the whole range so much as by reduced sensitivity to certain frequencies or a progressive reduction in hearing with higher frequencies. In some countries damage to the hearing of workers is now legally the responsibility of the employer. In their own financial interest as well as in the interest of the worker's health, some companies take audiograms of new workers who are likely to be exposed to high noise levels. The level at which noise has been demonstrated to interfere directly with performance is not very different from the level at which there is risk of damage. There may be indirect interference with work at much lower noise levels, in that such noise interferes with communication and can thereby increase the risk of accidents or lower morale. It is difficult to make any generalizations about the annoyance caused by noise, since this is very much a function of particular situations and of individual differences. For instance, noise made by other machines or departments is much more annoying than the same noise made by the worker's own machine or department, intermittent unpredictable noises are more disturbing than expected noises, and manifestly unnecessary noise is more provoking than inevitable noise. A group whose morale is high will often ignore noise, but if morale drops for quite other reasons, there may be complaints about noise. This effect is so common that in a situation where there is a sudden increase in complaints about noise in the absence of any obvious change in noise levels, it should be suspected that such complaints are an indication of more general dissatisfaction. There are various kinds of ear defenders providing protection against noise but they are not usually popular with workers even though they interfere much less with speech communication than one would expect. The most effective action against noise is to do something about the source: either change the mechanism—e.g., by substituting squeezing forces for impact forces—or insulate it from the workers by distance or by solid barriers. Unfortunately noise travels easily round corners and even through small gaps, and absorption is usually a direct function of the mass of the barrier.

The problem of speech communication is a highly complex separate topic best left to the specialist. For example, although almost all the energy in speech is in the band 100–1000 cycles per second, it is possible to remove all energy below 2000 cycles per second and still leave the speech intelligible. It should also be remembered that face-to-face communication does not depend entirely on hearing, since lip reading and gesture interpretation play important parts. On the other hand, the interference with conversation rather than the noise level as such may be the major source of annoyance or inefficiency in a particular situation.

The mechanisms of the *visual system* are not yet fully understood and understanding is not improved by the drawing of false analogies between the eye and the photographic camera. It is true that there is a lens, a focusing mechanism and a diaphragm (the iris), but the retina on which the image falls has nothing in common with a film except that both are photosensitive. The retina is not homogeneous, only a very small part of it (the central fovea) immediately opposite the lens registers clear images. Most of the retina is concerned with detection of movement in the periphery of vision, which is a cue for eye movement. The analysis and integration in space and time of retinal images is done in the brain. Thus the eye is best thought of as two mechanisms—a movement and change detector that controls eye movement via the brain, and a very narrow cylinder of clear vision, rather like a small searchlight, that stabs about in the environment picking up samples, which are put together with earlier samples in the brain to make up the three-dimensional "visual picture" of the environment. Colour vision is also restricted to the central area of the retina. Apart from iris changes, the level of sensitivity of the eye is controlled by highly complex changes in chemicals associated with receptors and with a switch, at low levels of illumination, from one kind of receptor-cones to another kind of receptor-rods. The rods provide only black and white vision.

Vision is the dominant sense in man. The study and design of the lighting environment is of corresponding importance. As in the case of sound, there is considerable confusion about standards, definitions and units of measurement. It is always convenient and unambiguous to adopt formally based physical units of measurement, but there is no escape from the fact that light, by definition, is energy to which the eye responds and, of course, eyes differ between individuals and respond differently to different frequencies and combinations of frequencies within the spectrum of electromagnetic radiation. However, the problems of photometry, or light measurement, are not too complex in principle since there are only four measures of importance. For each of these four measures there are a number of different terms and also a great variety of names depending on the physical units used. For example, *phot* is a measure of illuminance in *lumens per square centimetre* and *lux* is a measure of illuminance in *lumens per square metre*. The four measures are: (1) *luminous intensity*, which is the " power " of a light source considered as a point radiating in all directions; this is measured in *candelas* or *candle power* and is the basic arbitrary unit; (2) luminous flux, which is the flow of light related to a unit of solid angle measured in *lumens;* (3) *illumination* or *illuminance*, which is the amount of light reaching a surface measured in *lumens per unit area;* and (4) *brightness* or *luminance*, which is the amount of light reflected from a surface measured in *lamberts*. The measure used in specifying lighting levels and determined by " light meters " is the illumination level. Brightness meters are also available.

In relation to light sources there are two other measures of practical importance. One of these—*luminous efficiency*—represents the proportion of total energy that is available as useful light, but because of complications in defining exactly what is meant by " useful " it is usually expressed not as a percentage but in terms of lumens per watt. The other useful measure is the *colour temperature*. The exact definition depends on a knowledge of radiation characteristics of black bodies but it is roughly true that the higher the colour temperature the higher is the efficiency and the less yellow the light becomes.

Because of the adaptability of the eye it is very difficult to arrive at lighting standards that have any scientific validity. Many standards have been published, often stating slightly different illumination levels for different jobs or work situations, but these standards are arbitrary and are subject to regular modification—that is, quoted standards of lighting increase with each new publication. In Europe it is currently fashionable to stipulate illumination levels at bench height of 500–1000 lux. This is not to deny that required levels are governed by various factors of the job, such as size, contrast, time available and movement. A useful rule of thumb is that the illumination level should be 30 times higher than the level at which the task can just be done. However, it is worth repeating that there are no exact lighting standards and it is usually better to err on the side of too much light, provided glare can be avoided, if only because lighting costs are usually very small compared to human labour costs. Glare is simply unwanted light from an unshaded source or excessive reflection from some part of the environment.

Age is, in this context, an unusually consistent and yet often neglected human variable. The amount of light required for the same level of visual effectiveness doubles for an individual about every 13 years. The transfer of older and often highly skilled workers from visually exacting jobs can be obviated or delayed by better design of the lighting environment. This is not merely a matter of providing higher illumination levels. Aspects such as glare, contrast, direction and colour must be taken into account. When designing overall lighting schemes the relationship between natural and artificial light and the avoidance of too much uniformity also need to be considered. One aspect of this latter is the " phototropic effect ", in which the eyes have some tendency to move towards the brightest part of the visual field.

The study of colour must also begin by a definition of the variables and their measurement. There are three aspects of colour: hue, saturation and brightness. *Hue* is what is normally understood by the word " colour "; it is defined by the location of the peak wavelength in the energy distribution of the light under investigation. The extent to which the hue can be defined depends on the degree of saturation. A *saturated colour* is one in which most of the energy is restricted to a very narrow band of wavelengths. *Brightness* is the same thing as illumination level. The spectrum of visible colours is within the range 10^{-6}–10^{-7} m. The usual measures are the nanometre and the ångström unit, which are equivalent to 10^{-9} m and 10^{-10} m respectively. The visible spectrum from violet through blue, green and yellow to red extends from about 400 to 700 nanometres.

Hues are best defined by the C.I.E. chart,[1] which is based on the fact that any hue can be matched by a mixture of red, green and blue primaries. The proportion of green is plotted against the proportion of red, and the proportion of blue can, of course, be inferred by subtracting the sum of these co-ordinates from 100. These primary colours, which are additive (an equal mixture of all three gives white), are not the same as primary pigments, which are essentially subtractive (an equal mixture of yellow, blue-green and blue-red gives black). There are still many variations on the theory of how colour vision works but this is not of any great consequence to problems of work design. Unfortunately there is correspondingly little systematic knowledge of the use of colour and of attitudes to colour. It is established by experience that saturated colours are attention-gaining and advancing so that they can be used for signals and to make surfaces look nearer than they really are. Correspondingly a wall painted in an unsaturated colour appears further away and thus a space can be made to look larger. Another fact based on experience is that the larger the room the more it is possible to use bright saturated colours, and *vice versa*. The main use of colour is in improving total appearance, although colour codes for discrimination can be used on wall displays and on wires and pipes. Moreover, it is sometimes useful to use a contrasting colour to focus attention on a part of a machine—for example, the inside of a lid that should not be left open, the background of a particularly important scale, or the handle of a control. It must be remembered that colours change and eventually disappear as the level of illumination is reduced. This is a consequence of the mechanism of dark adaptation, a process that takes about half an hour to complete in an individual, so that sudden changes of illumination level, either within one place or as the worker moves from one place to another, can be dangerous.

General Principles

1. The distant senses, vision and hearing, have a wide range of adaptability but there are definite limits. Too much light and too much noise are dangerous to the mechanisms involved.

2. With rare exceptions, such as the intensity of light when looking at the sun or noise in severe thunderstorms, the dangerous levels only occur

---

[1] Devised by the International Commission on Illumination *(Commission internationale de l'éclairage)*.

in man-made situations. Thus man does not have the experience to take intuitive precautions and he will not avoid danger unless he is trained to do so.

3. Psychological effects of noise are difficult to predict because of the dominant influence of predisposing attitudes.

4. For skilled workers vibration and other pressure changes are an important source of information about machine operation.

5. Exact lighting standards for particular jobs are not available, obtainable, or even necessary because of the adaptability of the eye.

6. Recommended lighting levels must take account of the age of operators and must be increased as operators grow older.

7. Factors of importance in lighting design, other than illumination level, are contrast and glare.

8. Contrast boundaries are the basic variables in relation to ease in seeing. A dark object against a light background can be seen just as easily as a light object against a dark background.

9. Contrast can be provided by difference of hue as well as difference of brightness.

10. The uses of colour in relation to work design are essentially pragmatic.

Particular Cases

1. Vibration in metal cutting or drilling enables the skilled operator to adjust to optimum pressures. For example, if automatic feeding or feeding by remote control is substituted for direct manual control, tool life is likely to be reduced.

2. Vibration level also serves as an important pointer to the state of a mechanism—e.g., when a car mechanic investigates an engine.

3. Speech communication in noisy surroundings is best improved by the adoption of clearer diction on the part of the speaker and by a reduction of distance (to a matter of inches) between speaker and listener, rather than by speaking more loudly.

4. Noise caused by hard objects coming into contact can be lessened by reducing hardness—e.g., by using wooden or plastic containers for metal objects in place of metal containers.

5. Noise due to inherently noisy mechanisms can be reduced by substituting other mechanisms—e.g., electric motors for air motors.

6. Discrimination of small objects is best achieved by improving contrast rather than increasing brightness.

7. Fluorescent tubes are about 5 times as efficient as filament bulbs. The greater uniformity of the fluorescent tube can be an advantage in reducing glare or a disadvantage in reducing visual cues from shadows.

8. The use of individual lights for machines and work spaces encourages increased visual concentration on the working area.

9. Colour differences are best seen in light diffusing from a surface; surface faults are best seen in the distortion of light reflecting from the surface.

LIMITATIONS OF THE SENSE ORGANS

A. LIGHT MEASUREMENT UNITS

Quantity measured		Recommended unit*	Other units
Description	Name		
Brightness of point source	Luminous intensity	Candela	Candle power ($\equiv$)
Flow of light	Luminous flux	Lumen	—
Amount of light reaching surface	Illumination / Illuminance	Lux	Foot-candle ($\equiv$) Lumen/cm^2 (phot)
Amount of light re-emitted by surface	Brightness / Luminance	Lambert	Foot-lambert Candles/cm^2 (stilb)

WHO 20210

* Recommended by the International Organization for Standardization.

B. LIGHTING LEVELS

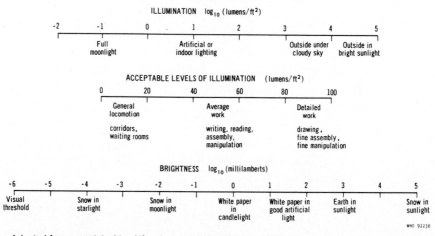

ILLUMINATION $\log_{10}$ (lumens/ft^2)

-2   -1   0   1   2   3   4   5

Full moonlight

Artificial or indoor lighting

Outside under cloudy sky

Outside in bright sunlight

ACCEPTABLE LEVELS OF ILLUMINATION   (lumens/ft^2)

0   20   40   60   80   100

General locomotion

corridors, waiting rooms

Average work

writing, reading, assembly, manipulation

Detailed work

drawing, fine assembly, fine manipulation

BRIGHTNESS $\log_{10}$ (millilamberts)

-6   -5   -4   -3   -2   -1   0   1   2   3   4   5

Visual threshold

Snow in starlight

Snow in moonlight

White paper in candlelight

White paper in good artificial light

Earth in sunlight

Snow in sunlight

WHO 91238

Adapted from an original in: *Life sciences data handbook*, published by the United States National Aeronautics and Space Administration, Washington, D. C.

C. LIGHTING OF THE WORK SPACE

GLARE

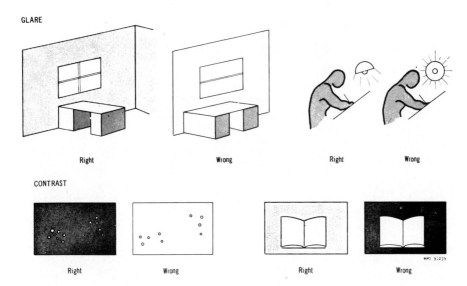

| Right | Wrong | Right | Wrong |

CONTRAST

| Right | Wrong | Right | Wrong |

D. COLOUR IDENTIFICATION

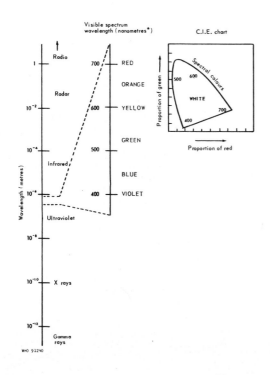

* 1 nanometre $= 10^{-9}$ metres.

The C. I. E. chart is based on the assumption that any colour can be produced by a mixture of the three primary colours, red, green and blue. Thus every colour can be determined by a point on the chart that indicates the percentage of green and the percentage of red. The percentage of blue can be calculated by subtracting the sum of these co-ordinates from 100.

E. INTENSITY OF DIFFERENT SOUNDS

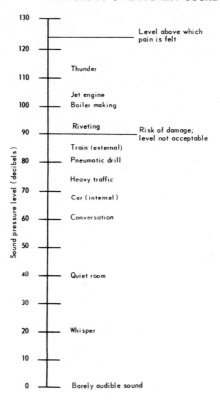

```
Sound pressure level (decibels)

130 ─
              ────────── Level above which
120 ─                    pain is felt

110 ─         Thunder

100 ─         Jet engine
              Boiler making

 90 ─         Riveting      ────── Risk of damage;
                                   level not acceptable
 80 ─         Train (external)
              Pneumatic drill

              Heavy traffic
 70 ─
              Car (internal)

 60 ─         Conversation

 50 ─

 40 ─         Quiet room

 30 ─

 20 ─         Whisper

 10 ─

  0 ─         Barely audible sound
```

WHO 91241

F. NOISE DAMAGE LEVELS

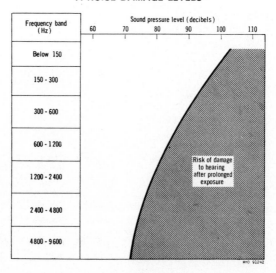

Frequency band (Hz)	Sound pressure level (decibels)
Below 150	
150 - 300	
300 - 600	
600 - 1 200	
1 200 - 2 400	Risk of damage to hearing after prolonged exposure
2 400 - 4 800	
4 800 - 9 600	

WHO 91242

G. VIBRATION AND MOTION DETECTION

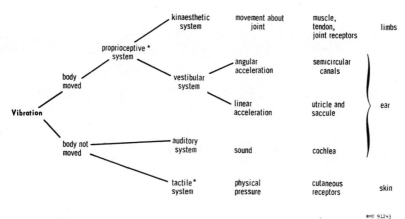

*Parts of somaesthetic system

WHO 91243

H. VIBRATION CONTROL

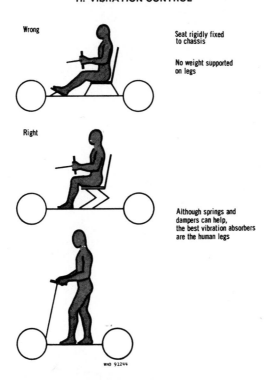

Wrong

Seat rigidly fixed to chassis

No weight supported on legs

Right

Although springs and dampers can help, the best vibration absorbers are the human legs

WHO 91244

Mild symptoms due to vibration: headache; misty vision; eye-ball tremor; changes in perception of distances; impairment of body balance; stiff neck and back muscles; digestive disorders; and low back pains.
Severe symptoms due to vibration: musculo-skeletal disorders; gastrointestinal disorders; and Raynaud's disease (vasospasm in fingertips when exposed to cold).

— 61 —

Reproduced by permission of Thorn Lighting Ltd., London, England.

CHAPTER 6

The Design of Controls

A control is a device that enables an operator to change the state of a mechanism. It converts the output of an operator into the input of a machine. Since controls are pieces of hardware they are often regarded as parts of a machine but, for design purposes, they can be more effectively considered as links between the machine and the operator. A control transmits information from the man to the machine and the starting point of design must be the output characteristics of the operator.

The range of these output devices is not large, consisting essentially of four limbs and a voice. Although the vocal output is versatile for man-to-man communication, it is of no use for communication between man and machine. Machines can store verbal information—e.g., on tapes and discs—but they cannot utilize it selectively. Thus, for communication with a machine the operator can only use his limbs. Since the basic output of the human limb is a force, one would expect pressure controls to be the most common, but in fact they are rare in comparison with displacement controls. A *pressure control* is one that effectively does not move at all, but the force or torque exerted on it influences the machine, while a *displacement control* is one for which the change of state of the machine is a function of the distance or angle through which the control has been moved. Pressure controls are usually more complex from the engineering design viewpoint and they suffer from the disadvantage that there is only one equilibrium position (no force exerted); other states can only be maintained while the force is being exerted. It has been pointed out in Chapter 2 that the human operator is not efficient in the generation of static forces and thus pressure controls are only useful in situations where the variable to be controlled and the corresponding control activation is changing continuously. This occurs in some tracking tasks. The pressure control has the advantage of providing immediate and unambiguous kinaesthetic feedback. The direct feedback from displacement controls is more complex since it depends on the control-restoring forces. These may not exist, in which case the control stays wherever it is put and feedback is a matter of sensing displacement. There may be spring-based restoring forces, in which case the displacement will be a function of the force exerted. If the control is subject to viscous damping then the velocity of movement will be a linear function of the force exerted, and if the control has a very high inertia (a large mass) then its acceleration will depend on the applied force. In practice every control has a mixture of all these effects with various weightings, and for sophisticated control design it is necessary to look at the equation of motion of a control as a function of applied force. However, when the operator is given information about the behaviour of the controlled variable he can quickly adapt to a great variety of control characteristics and thus the kinaesthetic feedback becomes secondary. This is not to underrate the importance of the " feel " of controls, and in fact one of the key features of control design is to enable controls to be used effectively without vision. To this end it is often useful to provide mechanical gates through which controls move and which provide pressure and displacement cues, which in turn enable the operator to know what he has done without looking at the control. This has been achieved very effectively, if accidentally, in many gear levers for motor cars, and the operation of such well-designed control mechanisms is a source of pleasure to the operator. All this applies generally, regardless of which limbs, joints and muscles are involved in the control actuation.

A second general problem consists in deciding which limb to use for which purpose. Broadly speaking, foot controls are best for the provision of powerful or continuous forces and hand controls are best for speed and precision. It is not possible for an operator to select between more than a small number of foot controls without the likelihood of confusion and errors. Probably four is the maximum number that can be used in the absence of the highest degree of practice and skill, such as that displayed by an organist, and one control per foot is the optimum. This assumes that the operator is seated and can maintain equilibrium without relying on his feet. The standing operator cannot use foot controls without extensive postural adaptation. This is likely to be uneconomic in speed and energy expenditure except for very infrequent use. The are wide individual differences in the ability to co-ordinate hand and foot control movements; thus a critical task that includes both will require an appropriate selection test. There are only a few kinds of foot control in common use: the pedal, the rotating pedal, the treadle, the rudder-bar and the rotating platform. Pedal design is largely an anthropometric problem of placement in relation to the floor and the seat; it is usually necessary to have an adjustable seat. The pedal must be provided with a non-slip surface and the restoring force should be large enough to support the weight of the foot. The rotating pedal and the treadle are really relics of the times when it was necessary to provide rotation by reciprocal foot activity, as in sewing machines and potters' wheels, but they have some advantages in providing flexibility. Usually either the right or the left foot or both feet can be used, and, if the axis of rotation is properly placed, restoring as well as establishing forces can be applied. The rudder-bar is superior for precise control by agonist/antagonist-type activities and is, moreover, the best control for maintaining body stability. The rotating platform is a device that is particularly useful for adjusting the orientation of machine or material in relation to the operator's hands—for example, when it is necessary to work on both sides of a chassis or all round a circular plate, as on the linking machine used in hosiery. It requires very good seat design since only an unusually small proportion of body-weight can rest on the feet.

The variety of hand controls is commensurate with the versatility of the hands. The steering wheel is the hand equivalent of the rudder-bar and has the corresponding advantage of achieving precision by balancing forces between the two hands. It makes possible a wide variety of position and type of hand-grip and operation by either hand or by both hands, and can also provide a large torque. In common with the bicycle, had it not been invented as a result of long experience, it would by now have been designed by the ergonomist. The handlebar is a specialized form of steering wheel with a reduced range of possible hand-grips. Cranks are the hand equivalent of rotating pedals, and since they are used to generate power rather than information it is usually relevant to inquire whether a pedal system might not be superior. The more precise hand controls, such as selectors, toggles and buttons, pose problems of identification and grouping rather than of individual design. (The design criteria for individual controls are now well understood and are applied commercially.) Problems of placement and grouping are common to displays and controls, which will be discussed in Chapter 8. Identification is increasingly a problem in view of current trends to centralize controls on consoles and pendants. With mechanical systems, where the controls are directly and obviously connected to the controlled member, there is no difficulty in determining what the control does. Electrical, hydraulic and pneumatic systems make the anatomical problems of control design easier but greatly increase the identification problems. If all the controls are put together and are separate from the controlled mechanisms then it is usually desirable to arrange them in some meaningful pattern that enables the operator to perceive the interrelationship. Often this can be effected by joining the control positions with flow lines indicating the movement of material, power or information. Labelling is frequently necessary but raises its own problems with regard to the meaning of words or symbols. Ease of interpretation may not be as widespread as the designer might assume, either between different countries or between specialists in different aspects of technology.

The relationship between the operator and the control cannot be studied in isolation from the relationship between the control and the controlled member, since the operator always gets information back from the mechanism, which he must relate to his past and proposed control actuations.

The simplest aspect of this relationship is in the directions of motion of controls. Particular populations have certain habitual expectations about how a control or a display or a control/display relationship should behave. These are called *population stereotypes*. The main *control stereotypes* are that an increase (in volume, speed, voltage, height, etc.) should be effected by a clockwise rotation, a movement downwards for a pedal and upwards or to the right or away from the body for a hand lever. These seem to be common to most populations but no doubt the habits are stronger for individuals and countries with long experience of machine operation. Whether a downward position of a toggle switch indicates " on " or " off " varies from country to country, and it is desirable to render this indication more explicit by the addition of a pilot light signalling " on " when illuminated.

The concept of *control order* is a useful generalization for the behaviour of mechanisms in response to control actuation. A zero order or displacement control is one in which the displacement of the controlled member (e.g., the gun sight or the spot on an oscilloscope) varies linearly with the displacement of the control. A first order or velocity control is one in which the velocity of the controlled member follows the displacement of the control—that is, the more the control is displaced the faster the controlled member moves. A second order or acceleration control is one in which the acceleration of the controlled member is directly influenced by the displacement of the control—the more the control is moved the faster the controlled member accelerates. Higher order controls are feasible mathematically and electronically but they are not used, since even a second order control requires skill from an operator that can only be acquired by very long training. This is not to say that the aim should be always to use zero order controls—a first order control, for example, has advantages in tracking steadily moving targets. In practice, natural laws often determine control orders and usually they are not purely of one order but a mixture. For example, a car accelerator is a mixed first/second order control of the distance moved by a car along a road. A rate aided control, which is a mixture of zero and first order, is often deliberately designed for the control of anti-aircraft guns. First order controls are sometimes found on machine tools; first and higher order controls occur increasingly in the processing industries. Although an operator can learn to control a system up to second order given sufficient practice, it is a demanding task. The higher the order of the system the greater is the concentration required and the less it is possible for the operator to " time share " with other tasks. There is also some evidence that stress from fatigue, fear or emergency conditions results in greater deterioration with higher order systems.

These aspects of design are usually described as *control dynamics* problems to distinguish them from anatomical problems. Thus, although it is necessary to match controls to the physical characteristics of the operators' limbs in terms of dimensions, angles, forces, etc. (functional anatomy), it is also necessary to match control systems to the limitations of the motor skills of the operator (dynamics). Motor skills are demonstrated by continuous patterns of output activity on the part of the operator. Throughout life the human being develops more and more complex motor skills. These begin as simple eye/hand or head/ear co-ordination activities and gradually build up through the skills of loco-motion to the complex highly integrated activities demonstrated by the games player, the process controller and the machine-tool operator. They are not easy to analyse or to understand because they are essentially unconscious, but they are nevertheless learnt and thus qualify to be called skills. In general what seems to happen is that any operator faced with a new task uses as basic building bricks the skills he already has, and these are gradually integrated into a new skill appropriate for the task. The essence of these skills is timing—the right thing is done at the right time by complex processes of anticipation and sequencing, and with control mainly by kinaesthetic feedback. This is why it is so important to take account of the dynamics or timing aspects of controls and the kind and quality of feedback provided by the controls. A control system that makes exacting but not impossible demands on the actual or potential motor skills of the operator is a pleasure to use. A control may be beautifully fitted to, say, the elbow height, the hand-grip dimensions and the available muscular forces, but it is still badly designed from the ergonomics standpoint if, when it is moved, uncontrollable activities are initiated within the mechanism or between the mechanism and the environment. The dynamics of control systems must be matched to the motor skills of the operator.

General Principles

1. Control design must be based on an understanding of what the operator is trying to control as well as how he can control it.

2. There are two general aspects of control design: *functional anatomy*—matching the controls to the musculature of the operator; and *dynamics*—matching the control systems to the motor skills and timing limitations of the operator.

3. There are two kinds of control in terms of what the operator does: *pressure controls*, on which he exerts a force, and *displacement controls*, with which he makes a movement.

4. There is a hierarchy of " control orders " that describes the time relationship between control movement and the resulting activity of the mechanism. The main ones are displacement (zero order), velocity (first order), and acceleration (second order). Most real controls are of mixed order.

5. The output activities of the operator must be allocated between hands and feet.

6. For maximum speed, selectivity and accuracy use the dominant hand.

7. For maximum power use one or both of the feet.

8. Before a control can be used it must be identified and selected.

9. Control axes of rotation should be designed to correspond to the position and mobility of the relevant joints of the body.

10. Accuracy of selection of controls can sometimes be usefully aided by using different shapes of knobs, thus providing tactile cues.

11. Controls that require flexion rather than extension of joints are usually to be preferred.

12. Controls that require extensive bodily activity—e.g., foot controls requiring heel pressure or any foot controls for the standing operator—are best avoided.

Particular Cases

1. If the reachable space on a console surrounding the seated operator is already fully utilized by displays and controls, further controls that are not frequently used can be placed in the roof space above his head. This is done in aircraft cockpits.

2. One common error in the positioning of foot controls is to place them so near each other that the large-footed and heavily booted man cannot press one at a time.

3. The design of foot controls for motor cars poses a particularly difficult anthropometric problem in that one design must suit every foot, from the small one in a high-heeled shoe to the large one in a heavy boot.

4. The pedal and gearing system of the bicycle has been investigated scientifically with the conclusion that optimum sizes and gearings have already been arrived at by trial and error as the bicycle evolved.

5. The speed, selectivity and force available to use a control will depend on which side of the body it is placed—e.g., a right-handed operator will only be able to generate with his left hand about three-quarters of the force he can generate with his right.

6. Gloves and boots need to be considered in relation to control design, particularly for cold or other hostile environments.

7. The need to operate a control should not interfere with the reception of new information related to the task—e.g., the crane driver must be able to operate his controls and see what is happening below at the same time.

8. Controls whose accidental activation may be dangerous should be designed with this in mind—e.g., guard rings for push buttons.

9. There are a few controls that are operated neither by the feet nor by the hands—e.g. the " knee-lift " control used on sewing machines.

DESIGN OF CONTROLS

A. TYPES OF CONTROL

		Use	Special design requirements
Button		In arrays for rapid selection between alternatives	To avoid slipping finger and accidental activation
Toggle		For definite, rarely used action involving only choice of two (normally on/off)	To avoid excessive finger pressure or nail damage
Selector		For more than two and less than ten choices	To avoid excessive wrist action make total movement less than 180°. Do not use simple circular shape
Knob		For continuous variables	Size depends mainly on resistance to motion. Use circular shape with serrated edge
Crank		When rotation through more than 360° is needed	Grip handle should turn freely on shaft
Lever		For higher forces or very definite activity	Identification of neutral or zero
Wheel		For precise activity involving large angles or rotation	Identification of particular positions Avoid slipping

WHO 91245

B. CRITERIA FOR CONTROL POSITIONS

ANATOMICAL			
Which limb?	Which joint?	When used	
		Force	Precision
Hand	Shoulder	High	Low
	Elbow	Medium	Medium
	Wrist/finger	Low	High
Foot	Ankle	High	Low
	Thigh/knee	Maximum	Minimum

PSYCHOLOGICAL	
Identification	Position
	Size
	Shape
	Colour
	Legend
Selectivity	Position in sequence
	Relative importance
	Frequency of use

WHO 91246

— 69 —

C. IDENTIFICATION OF CONTROLS

METHOD		WHEN USED
POSITION	EXAMPLE 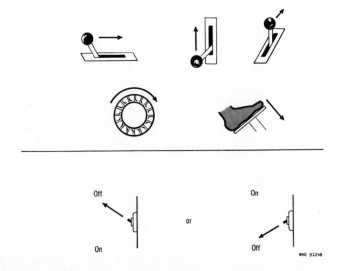	Best placed near to controlled function (mechanical device or dial) provided access is reasonable and high speed is not essential.
SIZE		Important in relation to high frequency of use (e.g.,space bar on typewriter) or in relation to operating force. Size should be increased with required force.
SHAPE		Useful when controls must be operated without visual attention.
COLOUR		Useful only as secondary cue or as warning unless there are many controls together. Even so number of colours used should not be more than five.
LEGEND	MAIN STAND-BY	Useful secondary cue. Should not be obscured when control is operated.

WHO 91247

D. CONTROL STEREOTYPES

WHO 91248

Arrows indicate direction of movement expected to produce an increase.
Standard position of switch indicating " on " or " off " differs from one country to another.

— 70 —

CHAPTER 7

The Design of Displays

A display is a part of the environment of the operator that provides him with information relevant to the task he is performing. The dial of a wrist watch is a display, as also are the driver's view through the windscreen of a car and the piece of paper on which is written information for a production manager about the state of his factory. These examples illustrate three types of display classified with respect to the environment. There are *real displays*, in which the changing environment provides the information directly to the sense organs of the operator—e.g., the windscreen,—and there are *artificial displays*, in which some hardware device intervenes between the real variable and the presentation of the information about it—e.g., the watch dial. There are *static displays*, which contain information about the task that does not change in time—e.g., the production date on the piece of paper—and there are *dynamic displays*, which reflect changes in time of the relevant variables—e.g., the windscreen and the watch. A real display is invariably dynamic but an artificial display can be either static or dynamic.

It will be noticed that the display only has meaning in relation to the operator and the task. Thus all problems of display design must be considered from the ergonomics point of view, in relation to the needs of the operator. There are, of course, engineering problems of artificial displays that centre on the physical parameters of the task, the cost and complexity of various sensing devices, modes of presentation, etc., but this need not concern us here except to note that, in practice, a display design must constitute some compromise between the engineering aspects and the ergonomic aspects. Thus, an artificial display is not solely a source of information for the operator, but nor is it merely part of a machine; it is best regarded as a link between a machine and an operator, and both sides need to be considered when it is designed. An operator carrying out a task must have some mental model of what he is trying to do, how best he can manipulate the physical world to achieve his objective, and how far he has progressed at any instant towards this end. This model is based on education, training and past experience of similar situations, and it is continuously updated by information from the environment—that is, through the displays and the sense organs. It follows that if we wish to design displays in relation to the needs of the operator such design must be based on a knowledge of the model that the operator is using and the properties of the sense organs that are needed to provide information to keep the model up to date with the current environment.

Unfortunately our knowledge of mental models, their acquisition and manipulation, is at present very limited. The psychology of human skill is a topic on which much more research is needed. The operator skills relevant in this context are the perceptual skills. These can be considered as a hierarchy of models of reality, the simplest of which are pictorial—that is, the model corresponds fairly directly to the real world. But these models become increasingly symbolic and abstract—e.g., a cat can be thought of as a real cat, as a drawing of a cat or as the word " cat ". To make the problem even more complicated an operator can shift up and down the levels of abstractness and symbolism during the performance of one task. For example, a maintenance operator looking for a fault in an electronic information-handling system might begin by asking the question: " Is the fault in the information sensing, in the store, in the data-handling unit, in the print-out, etc. ? ". Here he is using a broad functional model of the system for which the appropriate physical realization is a functional block diagram. Subsequently, he might ask himself: " Is the fault in an amplifier, a transducer, a magnetic core unit, etc. ? ". Here he is using an operational model of the system for which a physical unit block diagram is appropriate. Then comes the question: " Is the fault in a transistor, capacitor, resistance, etc. ? ". A circuit diagram is the appropriate model in this case. Finally he will ask the question: " Is it this real component or that real component? ". Here he is using a pictorial model that corresponds directly to the real layout of the components. This operator has shifted through at least four models, and ideally different displays should be designed to aid him at each level of his thinking. In practice, if the fault-finding system is well designed, he will be so aided. There will be static displays in the form of a machine manual containing the three kinds of diagram mentioned above; there will also be a real display of the component layout; and, finally, dynamic displays will be available consisting of light patterns on the console of the information-handling system plus meters and oscilloscopes that provide additional discrete items of information as he manipulates the appropriate probes or switches.

The problems described above are generally subsumed under the title " *coding* " or " *encoding* " problems, which form the core of all display design. Broadly the aim is to answer the question: " What information does the operator need and how is it best presented so that it matches the way he is thinking about his task? ". If this can be achieved successfully the operator is not obliged to do very much recoding of the information himself before he can use it. It emerges also that good display design can be more than the presentation of currently required information: it can be used to steer the operator into developing the right kinds of model when thinking about his task.

Another interesting consequence of the above point of view is that, since the kind of model used may change as the level of skill of the operator increases, the best display design may not be the one that is most easily comprehended at first sight. The coding problem can also be approached from a physical rather than a psychological point of view in terms of another classification of displays: pictorial, qualitative and quantitative. For simplicity, examples are given from visual displays but this division is not restricted to the visual medium.

A *pictorial display* consists of some level of direct representation of the real situation—for example, a spot moving across a map representing position, or a tiny model aircraft and an associated line, with the model and line moving in consonance with the real aircraft and the horizon.

A *qualitative display* indicates a general situation rather than a numerical description of it. This is often quite adequate for the needs of the operator. For example, a light showing when the oil pressure is too low is satisfactory for most car drivers, rather than a gauge that indicates the pressure in units.

A *quantitative display* presents a number denoting the value of some variable in the situation. There are two main types—the moving-pointer/ fixed-scale display and the digital display. The fixed-pointer/moving-scale display—a hybrid of the two—is not now commonly used because of the confusing nature of the relationship between the scale direction, the scale movement and the pointer.

This last point is one aspect of the more general problem of population stereotypes, already mentioned in relation to controls. For displays, the expected patterns in European and North American populations are that scales should increase from down to up, from left to right, and clockwise. These probably originate respectively with the real increase from down to up as items are stacked on top of each other, the reading convention of starting from the left, and the convention of the clock dial, which is the most widely known and first-learnt artificial display. Presumably this last was originally made to correspond with the sun-dial, since in the northern hemisphere the shadow moves clockwise round the dial. Although these conventions with regard to dials originated in the northern temperate zone it is likely that they are now common to all populations of the world.

These conventions have some relationship to the question of what happens between movements in artificial displays and movements in real displays. Again these should be made to correspond when possible. In most working situations the operator uses a mixed display, part real and part artificial, and since the real display cannot be altered it is important to design the artificial one so that it corresponds to the real one as far as possible in pattern and movement but not necessarily in complexity. In fact the basic difficulty about real displays is that they are very " noisy "— that is they contain many data that are not relevant to the task—and they are also excessively redundant—that is, there is much repetition and duplication. In the artificial display, on the other hand, noise and redundancy are at a minimum. Thus there is an inherent incompatibility between the two types of display, which, however, is only serious if it leads to the operator's having to change models when he switches between the real and the artificial. To put it another way, part of the art of display design is to ensure that there is a minimum of " attention shift " between the two displays. For example, the problem of the pilot flying an aircraft at high speed at a low altitude is that he must continuously shift between a noisy, redundant real world containing essentially analogue data seen through the windscreen and a pure, abstract, artificial display containing numerical data such as height and speed. Time is needed to shift the attention rather than merely to move and refocus the eyes, unless the pilot can develop a very complex hybrid model.

Occasionally it is useful to construct displays that record not the current state of the system but either the past or the future. These are known respectively as *storage displays* and *predictor displays*. There are also *quickened displays*, in which the state displayed contains some weighted derivative information, and *unburdened displays*, which contain weighted integrative information.

The coding processes having been dealt with as comprehensively as possible, there still remain the problems of matching the information to the available sense organs.

Again we can start from first principles, which in this case are relatively straightforward. There are only three important channels through which the operator receives information—the eyes, the ears and the sense of force and motion, or, to use more scientific terminology, the visual, auditory and kinaesthetic sensory channels. The task of the ergonomist is to allocate the required information between these channels according to their relative advantages and disadvantages and to ensure that for each channel the signals are above the threshold value. This is called the *psychophysics* of display design, as distinct from the encoding. It is not usually difficult or critical, since the sensory capacity of the human operator is very much greater than his decision-making capacity. This, of course, is basically why coding problems are given a higher' priority, but nevertheless there is no excuse for gross error in the psycho-physics of display design.

The commonest mistake is to assume that all the information must go through the eyes. It is true that in man the visual sense is dominant, but nevertheless the other sensory channels do have their unique advantages. In particular, design aimed at achieving optimum use of kinaesthesis is important, as mentioned already in relation to control design. The enormous advantage of kinaesthetic information is that it can be easily received and acted upon without conscious attention. There are also other advantages, including the avoidance of machine lags. One of the snags about visual and auditory information is that, when the operator has taken action, he usually has to wait for the machine or other hardware to react accordingly, before he gets back information about what he has done. This does not always apply to the kinaesthetic channel, where direct feedback from the feel of what has been done is immediate. One further minor advantage is that the kinaesthetic reaction time is marginally faster than either visual or auditory reaction times. The auditory channel has two main advantages: rhythm detection is particularly good and reception is multidirectional. One of the obvious but often neglected aspects of vision is that one cannot see something without looking at it. There is no corresponding limitation in hearing.

In spite of these points it remains true that vision is the main channel for human operators, fundamentally because they rely mainly on visual images or models in interpreting the environment. Apart from the dimensions of space the eye can accept data in colour dimensions of hue, brightness and saturation and it can also detect movement. The selectivity of the eye has minor disadvantages for detection but it does make it possible to present many different streams of data at one time. In practice this results in the multidial console. The relative pattern of the dials is limited by ease of seeing and the perceptual limits imposed by the need for structure—for example, if more than five dials or lights in a row or column are needed there must be divisions into sets of five or less. Different dials or lights can be identified by position, size, colour, shape, and, as a last resort, by label. Dials or lights may be grouped according to priority of use, frequency of use, or sequence of use.

In communication theory terms, the psycho-physics problem is one of adequate signal/noise ratio—that is, the information to be conveyed to the operator must be in a signal that is properly distinct from the inevitable noise. " Noise " in this sense can be visual or kinaesthetic, as well as auditory. For visual displays, therefore, scales, pointers and numbers of adequate size and contrast at an appropriate illumination level and in an appropriate position are required. For all these variables there is an optimum, which is not a maximum—too large sizes will take up too much space, too much contrast may cause too great an incompatibility with the surroundings, and too much light may result in glare and excessive adaptation. For most purposes the optimum is a broad region rather than a very precise level, so that common sense is sufficient, but for the occasions when more sophisticated design is needed there exist psycho-physical data on acuity, illumination, contrast, distance, movement, and their interrelationships.

For auditory displays, such as warning devices, similar data are available. The question of communication by speech is a very complex subject dealt with in specialist textbooks. For kinaesthetic displays the main noise problem is one of vibration, although, of course, changes in vibration can be part of the signal, as for example when cutting metal with a machine tool.

Although coding and psycho-physics problems have been described separately, in practice they interact considerably and final solutions are usually arrived at by a complex and variable series of decisions to be described in more detail in the next chapter.

General Principles

1. Display design must be based on a clear definition of the task and on an understanding of the way in which the particular kind of operator performs it.

2. There are two general aspects of display design: encoding—matching the display to the perceptual models used by the operator; and psycho-physics—matching the display to the sense organs of the operator.

3. There are three kinds of display: pictorial, qualitative and quantitative. Quantitative displays are only used when numbers are essential to the task.

4. In most tasks the operator receives information both from a real display and from an artificial display. The artificial display must be so designed as to be compatible with the real display in terms of patterns and relative movements.

5. When the information to be presented artificially has been assessed, it must be allocated:

(a) between the three sensory channels: visual, auditory and kinaesthetic; and

(b) between dynamic and static displays.

6. For maximum speed and minimum attention, use the kinaesthetic channel.

7. For maximum attention, use the auditory channel.

8. For maximum precision and agreement between operators, use the visual channel.

Particular Cases

1. Numerical and verbal data are essentially abstract and should be used mainly for intellectual tasks, such as the allocation of resources and fault-finding.

2. Static displays are an important part of integrated display design. Maps, legend plates, schematic diagrams, machine manuals, etc., come into this category.

3. Fixed-pointer/moving-scale dials are confusing and are best restricted to dials that must be read with great accuracy.

4. Use a digital display if the number is the only information the operator needs. If he requires the rate of change as well as the current value then use a moving-pointer/fixed scale.

5. Do not use scales that increase anticlockwise, from right to left or from up to down.

6. Do not use more than two pointers on one dial. It is much preferable to use only one and to extend the range by using a mixed counter plus pointer/scale display.

––––––––––

DESIGN OF DISPLAYS

A. TYPES OF DISPLAY

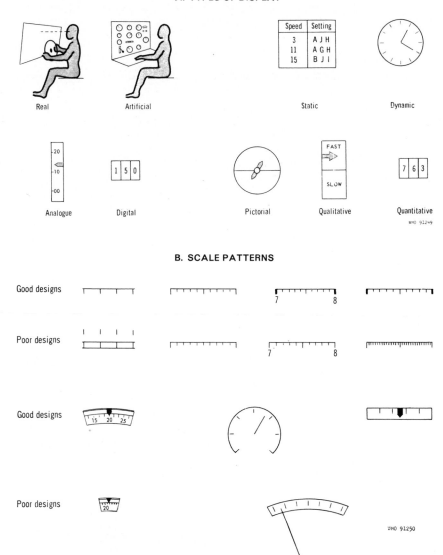

Speed	Setting
3	A J H
11	A G H
15	B J I

Real Artificial Static Dynamic

Analogue Digital Pictorial Qualitative Quantitative

WHO 91249

B. SCALE PATTERNS

Good designs

Poor designs

Good designs

Poor designs

WHO 91250

C. DIAL PATTERNS

Superior design

Poor designs

Reasonable alternatives

WHO 20209

D. DISPLAY STEREOTYPES

Expected

Unexpected

WHO 91252

— 80 —

E. COMBINED USE OF TRADITIONAL DIALS AND CHARTS AND CATHODE RAY TUBE
DISPLAYS AT WYLFA NUCLEAR POWER STATION, NORTH WALES

Reproduced by permission of British Nuclear Design and Construction Limited, Leicester, England.

CHAPTER 8

Man/Machine Information Exchange

The man/machine interface is an imaginary plane across which information is exchanged between the operator and the machine. Information is conveyed from the machine to the man by the display elements of the interface, and from the man to the machine by the control elements of the interface. The separate problems of displays and controls have already been discussed, but there are also the more general aspects of man/machine information exchange.

This is inherently a difficult design problem that cannot be left to chance, to common sense or to tradition. The two components are entirely different in their basic properties. The machine is fast, accurate, powerful and inflexible; the man is slow, subject to error, relatively weak and yet highly versatile. The nature of these properties explains why the man/machine combination is so useful, but only if these two fundamentally mismatched units can be efficiently linked together. In the earlier phases of the technological revolution the problem was not so important, since the machines' limitations and speed of performance were such that the design problems were in the hardware and the man could usually use his adaptability to compensate for any interface design weaknesses. However, given more efficient machines, the weaknesses of the total system are increasingly found at the interface. There has also been a change of attitude, in that when machines were novel men were prepared to adapt to them, but now it is expected, quite properly, that the machine should be adapted to suit the man by better interface design. This becomes an increasingly difficult design problem as the information exchange gets more and more abstract. For example, intuitive ideas about interfaces are adequate when guiding a plough through the earth, where the only display is the real one of earth movement and plough movement and the only controls are the plough handles. At the other extreme are problems of control rooms, where many abstract parameters are displayed and various combinations of them are influenced by a wide variety of controls.

The more abstract the situation becomes the more important it is to conform to the basic principles of interface design. Population stereotypes have already been mentioned for both displays and controls, but there are also stereotype aspects of *control/display relationships*. Broadly these follow from the relationships established in natural situations. Movement should follow the line of a force: if a control moves upwards, left or away from the body, then the corresponding display should move respectively upwards, left or away from the body. Relationships between rotary movements of controls and linear movements of displays usually follow the logic of the simplest mechanical connexion. For example, right-hand threads are such that a clockwise rotation causes an away movement and this is the expected relationship even when there is no mechanical connexion between the control and the display. Stereotypes for rotary controls adjacent to linear displays follow simple ratchet or pulley/belt-type movements. Relationships between display movements at right angles to control movements are usually ambiguous because there are a variety of equally valid ways in which a mechanical connexion could be envisaged. These should be avoided. Unexpected display/control relationships may result in longer response times, greater errors, longer training times and high fatigue rates. Even when a new relationship has been learnt for a new situation there may be a regression to earlier habits when an emergency arises. For example, if the brake and accelerator pedals of a car were interchanged, it would be perfectly possible, with extra care, to drive such a car. However, if the car had to be stopped suddenly there would be a tendency to press the accelerator pedal which would, of course, have the contrary effect. There have been many serious accidents resulting from situations analogous to this.

The most confusing machines are those in which stereotypes and reversed stereotypes are present on the same machine so that when the operator moves from one control to another he has to change his stereotype. This can happen on machines such as multiaxis machine tools, where on some axes the tool is moved but on one axis the work is moved. For example, on a horizontal boring machine it is often true that the tool moves " upwards/downwards " and " in/out " but the work piece moves " left/right ". Movement is merely relative and it may be that the work rather than the base of the machine is used by the operator as a reference point.

It is reasonable to assume that if a particular mistake is made frequently either by one operator or by a number of operators, there is probably something wrong with the interface. The operator may not be getting adequate information or the information may not be coded in a way that encourages him to make the adaptive control movement.

This is not to suggest that the operator should be regarded as fundamentally unthinking, unskilled and irresponsible, and that the foolproof approach to interface design should always be adopted. On the other hand, neither should the designer take his task too lightly and assume that a skilled intelligent operator will be able to compensate for lack of skill in design. If a machine is to be used by the general public then it must be foolproof, but if it is designed for a specialized purpose and for use by skilled operatives then the objectives become more complicated and more debatable. The snag is that the criterion of simplicity and the criterion of versatility are often contradictory. For example, combining all the variables from a process into one display and providing the operator with one control is reducing the task to its simplest form but it does not leave the operator with very much freedom of action. If he is given displays representing all the variables in the situation and controls that affect all these variables, he will have greater versatility and greater motivation. Eventually he will achieve a better and more flexible machine performance. It is of no use to leave an operator in a system because of his intelligence and adaptability and then so to design the interface that he cannot exercise his abilities.

Unfortunately, the ability to produce a design that will give the operator freedom to do the right things but not the wrong ones still largely depends on flair and experience. The background to display and control design has been described; more will be said later about work design in general, but in the approach to interface design, only principles—and not fixed procedures—are available as a guide.

Taking the four main aspects—psychophysics, encoding, dynamics and functional anatomy—there is no optimum order in which these must always be dealt with for every interface problem. Since making decisions in any one area will restrict possibilities in all the others it is best to start with the one that corresponds to the most probable weakness in the information processing of the operator—that is, sensing, perceiving, timing and activation respectively. In power systems, such as machine tools and transport vehicles, the output side is likely to be the weakest, but in information systems, such as vigilance tasks and data-processing, the input side will probably be the most deficient. The designer can be reminded of all the factors that he should take into account by the provision of an appropriate check-list.

Given that almost all the power transmission occurs within the hardware, interface problems are almost entirely informational. The measurement of work or effort in functioning at an interface is therefore extremely difficult. Physiological measures of load are not appropriate, and even when used indirectly as measures of strain (e.g., muscle tension or sinus arrhythmia) it is not possible to make comparisons between systems. Psychological measures of information processing and level of skill (e.g., required channel capacity or precision of timing) are more obviously relevant, but at present they are unreliable and tedious. Even if it is possible to quantify the amount of information crossing the interface, this does not incorporate the critical encoding and dynamics variables. Informational measures of difficulty have been proposed and used for simple laboratory situations, but they have not been extended to cover even simple practical interfaces. At present, there is no satisfactory way of measuring how hard or at what level a man is working when he is operating at a machine interface.

This is tiresome not only from the practical point of view, but also from the theoretical, when attempts are made to validate or compare particular examples of interface design. The standard psychological measures of speed and accuracy can be used, of course, but it is difficult to incorporate such key variables as level of learning and degree of adaptability. This is probably the main reason why progress towards a general theory of interface design is so slow. However, it is worth remembering that a great deal can nevertheless be done to improve the design of interfaces beyond the level commonly found at present. Such systematic work usually gives greater rewards with the more complex systems now coming into use. Although the effects of good interface design on motivation and job satisfaction are even more difficult to measure it seems from anecdotal evidence that they are considerable.

General Principles

1. Man/machine information exchange is inherently difficult because of the very different advantages and limitations of the human being and the machine.

2. For this reason the weaknesses of man/machine systems can often be traced to inadequate interface design.

3. Population stereotypes in the form of control-display relationships mostly conform to the behaviour of real rigid systems or simple mechanical linkages.

4. Contravention of population stereotypes can lead to:
(a) longer response times;
(b) greater errors and rates of error;
(c) longer training times; and
(d) higher fatigue rates.

5. The foolproof objective of interface design is only valid for unskilled users.

6. The criterion of simplicity often conflicts with the criterion of versatility.

7. There are four general aspects of interface design: psychophysics, encoding, dynamics and functional anatomy. The order in which these are dealt with in a particular situation depends on where the operator is likely to have the most difficulty. Check-lists can be used as a reminder of the many factors to be considered.

8. There are no satisfactory general measures of skill and effort on the part of the human operator at an interface.

Particular Cases

1. For interfaces where speed and co-ordination of action are imperative it may be desirable to separate controls from displays, so that all controls are together. More usually, it is better to leave controls with associated displays and allow the operator greater mobility. This may be slightly more time-consuming, but any such disadvantage is more than offset by the reduced probability of error.

2. Interface panels should be vertical rather than horizontal, if only to prevent people putting things on top of them.

3. With regard to interfaces in transport systems—e.g., in driving a tank or in monitoring and controlling a ship's energy supplies and services—particular problems often arise from vibration and motion. Operation when in motion may be quite different from operation when stationary.

MAN/MACHINE INFORMATION EXCHANGE

A. THE MAN/MACHINE INTERFACE

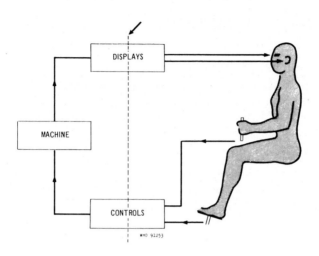

B. INFORMATION PROCESSING AND DESIGN

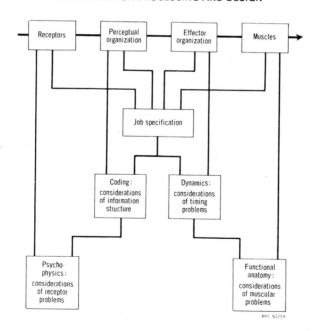

C. PATTERNS OF DISPLAYS OR CONTROLS

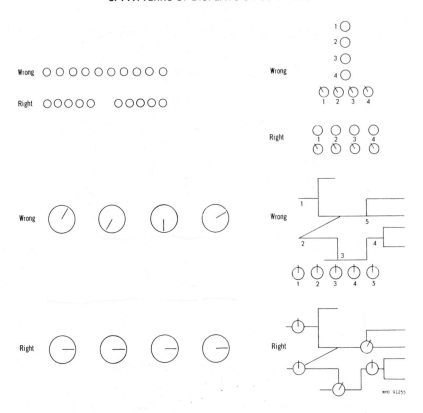

D. CONTROL/DISPLAY RELATIONSHIPS

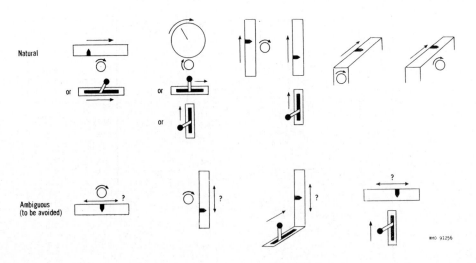

WHO 91256

— 88 —

E. CONVENTIONAL RAILWAY SIGNAL BOX WITH MECHANICALLY LINKED CONTROLS

Reproduced by permission of British Rail.

F. MODERN RAILWAY SIGNAL BOX WITH ELECTRICALLY LINKED CONTROLS

Reproduced by permission of British Rail.

CHAPTER 9

Temporal, Social, and Economic Conditions of Work

All human activity is subject to rhythmical variation. The circadian rhythms of the body have their counterpart in natural production rate, error rate, level of arousal and effort—although, of course, the length of time spent at work will also affect these activity patterns.

The *length of the working day* is surprisingly constant between different countries and industries. Moreover, it has changed much less than, say, the standard of living in the last 50 years. There is no simple answer to the question of how much time should be devoted to work. It is known that 60 hours a week is excessive to the point of being obviously uneconomic for production work. The optimum length of the working day is a function of many factors; various isolated bits of evidence suggest that the worker can adjust his rate of effort roughly to match the number of hours he knows that he will have to work. For example, piece workers working only half-days often have much higher hourly pay than those working full days. It can be predicted that a reduction of the working week by one or two hours will have little effect on weekly production. Certainly it is not valid to try to estimate production merely on the basis of the number of hours of work available.

Rest pauses are necessary in every occupation, the required frequency apparently being more a function of monotony than of intensity of work. This assumes that the operator is not physiologically stressed; if he is, then the necessary rest pauses can be computed from energy expenditure measures. More usually in modern industry the problems are psychological, to do with attention and motivation. If the work is unpaced the operator, given adequate incentives, will arrive at his own optimum. If the work is paced then rest pauses must be provided by deliberate design—for example, by allowing queues to form at the conveyor belt. A change of work can be just as beneficial as a rest. It is impossible to generalize about the frequency and length of rest pauses because of variations between jobs and between individuals. Study and prediction based on the measurement of production rates are notoriously difficult. For example, when total output drops, this is often due to a change in the pattern of work during the day, with a longer warming-up period and a more obvious and earlier fall-off towards the end of the day, rather than to a drop in the routine mid-period maximum rate.

Shift work has been studied on a large scale but the volume of knowledge accumulated is not commensurate with the amount of research that has been done. A wide variety of measurements of bodily activity—for example, temperature, metabolic rate, and glandular activity—follow a cyclical curve over 24 hours, normally with a maximum in the afternoon and a minimum in the early hours of the morning. When an operative goes on to night work this pattern is disturbed and, in the case of most individuals, will eventually reverse so that the maximum occurs during the night shift. There are large individual differences: in some individuals the rhythm changes phase in two or three days, in others this takes much longer, and in yet others there is no change at all. Direct measures of activity during shifts have not yielded consistent results. In terms of production per unit time it is difficult to demonstrate differences between night shifts and day shifts; the difference is probably less than 5% in most cases. The evidence about accident rates from different research workers is conflicting. So also is the evidence about absenteeism. Often absenteeism is less on the night shift than on the day shift, but it has been found to be greatest in the second week of continuous night shifts. Although physiological factors are obviously involved, the range of results and conclusions suggests that psychological and sociological factors are predominant. For example, hospital staff find it easier to adjust to continuous night shifts than do industrial workers, since daytime sleeping conditions are better for the former. Results also seem to be affected by factors such as whether or not shift work is traditional in the neighbourhood. This may be sociological, in the sense of community acceptance, and partly environmental, in the sense of availability of transport and recreation facilities outside normal hours. The situation is currently becoming more and more complicated with the introduction of new patterns of shift systems that vary enormously between industries and countries. It is clearly a problem to be studied at the community and area level rather than in isolated factories.

The *pattern of culture* is involved in this and all other problems of work. Such a pattern represents the sum total of behaviour, attitudes and values within a community. One of the difficulties that always arises in trying to make generalizations about human behaviour, or, more particularly, worker behaviour, is the effects of the culture pattern. The *group behaviour* of the operator may be difficult to define exactly but it is obviously important. Man is a social being interacting with many other individuals within and outside his working group. These factors have a strong influence on his efficiency as a worker.

It is not possible to understand the hierarchy of authority within an organization without some background knowledge of the social structure and class structure. The *class structure*, in the sense of criteria for forming and ordering groups in the community, affects the concepts of role and status of the worker in the factory. It influences personality development, attitudes to work and to change, and communication systems within the organization. The *social structure* of a community will be affected by conditions of work and by technological change, but the effects are by no means all one way. It is not possible to understand, for example, the reactions of workers to change without looking at their total environment.

Even incentive rates of pay depend for their effectiveness on the financial requirements of the individual. They work very well in communities aspiring to increase their standard of living and where the goods and services are available to achieve this higher standard. They do not work so well where such conditions do not obtain. Again, it is impossible to quantify these effects but, on the other hand, it is of no use to carry out controlled experiments in a factory with such variables as method and extent of payment unless these external factors are also taken into account.

The general problem of *incentives* has three aspects: financial, environmental, and social. The fixing of rates of pay is obviously a complex business, governed by considerations ranging from the ethical requirements that the worker should earn enough to keep him alive and in reasonable health to competitive factors associated with the general demand for labour. Closely allied to the ethical considerations is the straightforward question of the necessity for adequate diet if work is to be maintained, as well as the often neglected aspect of the cost of replacement and retraining. Human operatives are not expendable either ethically or economically. At the other end of the scale there is the concern with fixing a limit above which the rate of pay for a job cannot go because of the need for comparability with the rates for other jobs and for maintaining labour costs at a tolerable level within total production costs. Further incentives must then be based on what are often called environmental factors, which include most of the factors described in detail in Chapter 13. Given that the pay is optimal and the job design is optimal, the only remaining incentives are in the social field. These include the status of the job in the factory and in the community, the satisfaction derived from working in a group, and other features that contribute to the overall state described by the word " morale ".

Morale is usually thought of as a state pertaining to some groups or organizations, although it may also be assessed in individuals and the results averaged in some way to obtain an idea of the group state. Attempts to measure or, rather, to infer morale can be made directly from productivity and quality levels, indirectly from absenteeism and accidents rates, and analytically by the use of attitude surveys.

One reason for making such attempts at measurement is that this is one of the intuitive tasks of the good leader; he must be sensitive to a change of morale in his group and try to do something about it when it falls. *Group leadership* is an important feature of any organization. There are many different styles of leadership, all of which can be summarized along the two dimensions active-passive and autocratic-democratic. The optimum style is a function of the personalities and the situation concerned, since each style has its advantages and limitations. The skilled leader can change his style to suit the problem or the state of the group. This he must do mainly on the basis of experience, but he can probably improve his skill by awareness of what he is doing and by an eclectic approach to the concept of the organization or the factory.

The factory can usefully be regarded at various times and from various points of view as an economic unit, a social unit, and a cybernetic unit. These different *perspectives of the factory* each lead to their particular insights, and no one concept is sufficient for all circumstances. The basic view, of course, is that the factory is an economic unit. On the other hand, the concept of the factory as existing solely to make money is as limited as the concept of the operator working only to make money. A factory is also a social unit, where some activities can only be understood in the context of groups of people who come together to achieve a common purpose. Another way of looking at a factory is to regard it as a cybernetic unit—an elaborate adaptive system attempting to maintain a continued existence and also stability in the face of continuously changing external pressures and circumstances. Its adaptability, like that of every other organism, is a function of its intelligence, in both the psychological and informational senses of the term, and also of its speed and variety of response.

General Principles

1. Human activity both externally and internally is fundamentally cyclical rather than uniform. Many rhythms, including those connected with work, are locked to the 24-hour sun cycle.

2. The optimum length of the working day or working week has not been determined with exactitude.

3. Variations of up to 5% in the length of the working day or working week have little or no effect on the total amount of work performed.

4. Rest pauses are essential in all work. For mental work or light physical work their duration and frequency should increase with the monotony of the work done.

5. Shift work affects bodily rhythms but it has not been shown to affect measures of performance to any great extent.

6. Behaviour at work is affected by patterns of culture, social structure and group behaviour.

7. There are three kinds of incentive: financial, environmental and social.

8. Morale, although important, is not easy to define or to measure.

9. The style of group leadership adopted can be varied to suit particular personalities and problems.

10. An understanding of the total behaviour of an organization requires a variety of approaches and concepts.

Particular Cases

1. Attempts in England to recruit married women as workers by introducing a special early evening shift have proved very satisfactory in terms of rate of work but less so in terms of reliability.

2. The need to increase machine utilization can sometimes adequately be met by employing workers on early and late day shifts, but not on night shifts.

3. Operators who have to account for all their working time—e.g., in a garage where many cars are being repaired and serviced—usually find this difficult to do unless they are allowed a certain proportion for rest pauses.

4. Training for leadership has in the past depended heavily on the development of isolating cultural patterns and attitudes rather than on skills as such.

5. Experiments on evaluating redesigned machines can be nullified because the operatives aim unconsciously to achieve standard piecework earnings.

6. In the history of direct competition—e.g., in warfare or in games—it has often been demonstrated that superior morale on one side can offset very large material advantages possessed by the other.

TEMPORAL, SOCIAL, AND ECONOMIC
CONDITIONS OF WORK

A. LENGTH OF WORKING WEEK

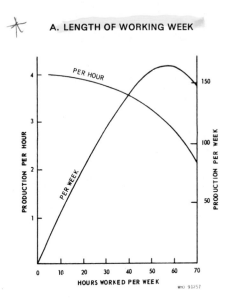

Since every operator reduces the rate of work as the length of the working week increases, and *vice versa*, there is an optimum length for a given operator, job, and current attitude. The optimum will change with each of these parameters. The above graph gives an approximate indication of average behaviour.

B. PATTERN OF DAILY WORKING ACTIVITY

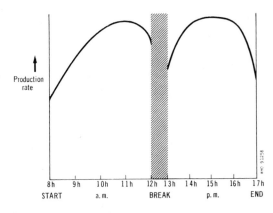

The precise pattern depends on the kind of work, length of the working day, and whether or not rest pauses are permitted, but the above graph represents fairly typical changes in the production rate when the rate is dependent on operator effort.

C. CIRCADIAN RHYTHM

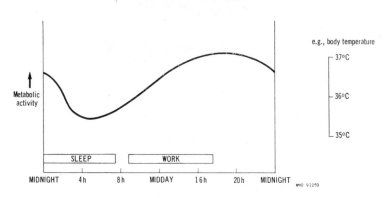

WHO 91259

Typical average activity curve for a day worker as measured by such physiological variables as body temperature, urine volume and content, and blood hormones. Psychological measures such as reaction time follow the same pattern, but usually lag slightly behind physiological changes. A change of working hours will disturb the pattern for several days.

D. SHIFT SYSTEMS

TYPICAL CONTINUOUS OPERATION SYSTEM
Schedule (2- to 4-day shifts)

Team A	Early	Late	Night	
Team B	Late	Night		Early
Team C	Night		Early	Late
Team D		Early	Late	Night

TYPICAL WEEK-END SHUT-DOWN SYSTEM

	Monday	Tuesday	Wednesday	Thursday	Friday	Saturday	Sunday
Team A	Early	Early	Early	Early	Early	Early	
Team B	Late	Late	Late	Late	Late		
Team C	Night	Night	Night	Night			

WHO 91260

Based on a 42-hour working week.
For the typical continuous operation system, 4 teams are required; the period on one shift being 2-4 days.
For the typical week-end shut-down system, 3 teams are required; the pattern repeats itself every 3 weeks.

E. SOURCES OF PRESSURES AFFECTING THE INDIVIDUAL

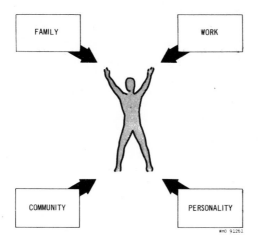

The reactions of the worker cannot be understood or predicted solely on the basis of what happens at work.

F. TYPES OF GROUP LEADERSHIP

	TYPE	CHARACTERISTICS	ADVANTAGES AND LIMITATIONS
DEMOCRATIC	Active	"Consensus government". Views of all members are sought and taken into account	Tedious and apparently wasteful, self-sustaining and satisfying, providing membership is not static
	Passive	"Laissez-faire". Every member pursues his own objectives and expects help from other members	Objectives become diffuse. Comfortable for members until internal or external pressures force changes
AUTOCRATIC	Active	Visionary leadership. Objectives and methods set by leader, members expected to conform.	Highly productive in short terms but not self-sustaining
	Passive	Any change or innovation is automatically treated as undesirable	Reliable and predictable but not adaptive

WHO 91262

G. MEASUREMENT OF CHANGES IN MORALE

Direct and immediate measures	Productivity
	Quality
Indirect and longer term measures	Absenteeism including strikes and unpunctuality
	Accident rates
Positive investigations	Attitude surveys

WHO 91263

H. PERSPECTIVES OF THE FACTORY

ECONOMIC UNIT	A unit attempting to provide a proper return on the capital invested

SOCIAL UNIT	A group or a large number of groups of people who come together to achieve a common purpose

CYBERNETIC UNIT	A dynamic unit attempting to continue to survive by adaptation to external events

WHO 91264

CHAPTER 10

Age, Fatigue, Vigilance, and Accidents

In any developed country nearly half the working population is likely to be over 45 years of age. The proportion of those aged over 60 has increased throughout the century from less than 10% in 1900 to between 15% and 20% today, and is likely to stabilize by the end of this century at about 25%. This, of course, is in marked contrast to the situation in developing countries, where half the population is likely to be under 20. It is not known how far climatic and other stresses result in premature aging, although even in temperate zones the correlation between chronological age and biological age is a fairly loose one.

There is a general decrease in the efficiency of the sense organs with age. In the case of the eyes, the elasticity of the lens decreases from early childhood onwards, but changes in acuity, width of visual field, perception of depth, and colour perception only become of importance in most jobs for workers over 40 years of age. Nevertheless, there are an increasing number of jobs, such as, for example, micro-assembly, in which the visual acuity requirements are such that only persons under 30 can suitably perform them. One of the most consistent of all aging effects is the decrease in the capacity for dark adaptation, which means that the older the worker the higher is the required level of lighting. Hearing losses with age occur most usually at higher frequencies. However, a marked hearing loss for frequencies greater than 15 000 cycles per second is not of importance in work. For communication purposes there is some compensation in that, although the absolute limit of hearing has changed considerably, the ability to hear selectively against a background of other noises is less affected. Direct evidence on changes in touch and movement sensitivity with age is sparse, but the effect may be considerable, since it seems that older people rely more on visual cues than on kinaesthesis.

Deterioration of the capacity for physical performance with age is smaller than might be expected, except in the case of intensive effort, which is restricted by a decline in both maximum pulse rate and muscle strength as measured by hand-grip and lifting capacity. Nevertheless, there is no evidence that people relinquish work merely because it is heavy, but there is some evidence that the tolerance for certain strenuous actions, such as excessive bending, is decreased.

The effects of heavy work may be confounded by the fact that light work is often fast work, which is more stressful for older people. In broad terms, from the age of 20 to the age of 60, muscular strength drops by about 25%, while sensorimotor performance declines by about 60% in relation to maximum capability. There is little reduction in the capacity for immediate memorization, but information is retained for an increasingly shorter time, and items of knowledge stored temporarily in the memory are more easily disturbed. The extreme case of this occurs in senility, when the very old person cannot remember things for long enough to piece complex situations together and sometimes forgets the purpose of actions in the middle of performing them. Less obvious instances of similar effects might be serious in some highly skilled work. Older people, of course, accumulate experience and also develop adaptive strategies in relation to their limitations that may or may not be successful, depending on the circumstances. For instance, the generally decreased information-handling capacity results in a greater tendency to ignore events, a narrowing of interests, and a restriction of activities; this may result in increased reliability, increased concentration, and decreased versatility. Older people may try to maintain the same speed as younger workers, which will result in more numerous errors, or they may reduce speed more than is required to compensate and thus decrease the number of errors. They will cope with a task by looking at it in smaller units, which may lead to failure or, on the other hand, to steadier progression.

This may appear to be a depressing picture but there are many redeeming features. Memories and varieties of experience increase and, as already noted, some people age much faster than others. Age is only one factor in individual differences and is not the dominant one—for example, in spite of the decline in strength with age some 60-year-olds are stronger than many 20-year-olds. Moreover, activity plays a very favourable role in the maintenance of physical, psychomotor and mental capacity. The effects of age are very similar to those of oxygen deprivation, and in many respects resemble the effects of fatigue.

The difficulty about the concept of fatigue is that it covers a wide range of phenomena and yet does not have an exact definition. Fatigue may be physiological (i.e., physical or chemical) or psychological (mental or functional); it may be objective (resulting in performance changes) or subjective (giving rise to changes in feeling and awareness). In general terms, fatigue can only be defined as an effect that emerges during continued performance.

Physiological fatigue is obviously connected with the supply of energy to the muscles and with the removal of waste products. If the blood flow is restricted then pain is induced and performance falls. This may be due to swelling caused by an accumulation of lactic acid with a consequent increase in pressure within the muscle sheath, or it may be caused by primary oxygen deprivation interfering with the biochemical reactions required to maintain chemical equilibrium. The characteristics of this kind of fatigue are nicely illustrated by an ergograph. This is a device providing a constant force against which a muscle can work and a method of recording the activity. An example of an ergograph in its simplest form is a weight attached by a string running over a pulley to a finger, the movement of the pulley or weight being recorded when the subject is asked to move the finger to and fro. It is found that if contractions are infrequent no fatigue occurs. Within a large range, the total amount of possible work is greater for smaller loads and for a decreased speed of work. The normal decrement in performance can be reversed temporarily by reducing the load. The amount of time needed for the finger to recover from fatigue increases rapidly with the amount of work—e.g., it may take half an hour after 30 contractions and 2 hours after 60 contractions. If the blood flow is restricted by a tourniquet on the arm then the fatigue appears very quickly. All these specific effects are highly consistent and easily understandable, but there are other more general effects. For instance, fatigue in one finger will spread to other fingers. Most difficult to explain of all is the fact that if the subject is told that the weight is now reduced, but in fact it remains unchanged, the effect is more or less the same as if the weight actually had been reduced. Thus even this, the apparently simplest kind of fatigue, has a psychological component.

Psychological fatigue in the sense of feelings and opinions (weariness) is notoriously inconsistent and subject to enormous fluctuations with slight changes of conditions and, more particularly, changes of morale. (A case in point is the obvious difference in apparent fatigue between the winners and the losers in any contest.) There are other subjective symptoms of fatigue that are not felt simply as tiredness, such as increased irritability or increased withdrawal, increased awareness of irrelevant stimuli such as hunger, or postural discomfort, and so on. Thus, it is evident that this kind of fatigue is inextricably bound up with questions of morale, motivation, monotony, general health, and so on, as well as with the immediate effects of the amount of work done and the rate of working. Unfortunately this is also true, if less obviously so, for fatigue in the sense of performance decrement. In fact, simple performance decrement is one of the later manifestations of fatigue and only occurs in sequential tasks where speed and accuracy are relatively independent. More usually there is some complex trade-off of speed and accuracy, with errors increasing suddenly or erratically while speed remains constant, or errors remaining constant and speed being subject to apparently random fluctuations. Even before this occurs there are more subtle effects that can be detected, such as increased variability of performance (even though, on average, there is no change in rate) and the disorganization of performance into smaller units. All this makes the subject very difficult to investigate under controlled conditions. For example, there have been many investigations in which groups of performance tests (e.g., intelligence tests, reaction times, tracking skill, and manipulative tasks) have been given at intervals during long periods of work or of sleep deprivation, and often there are no significant performance changes although the subjects must be getting fatigued by any definition. It seems that, for a short time at least, the operator can always completely mask the effects of fatigue by increased effort. There are now signs of greater success in measuring fatigue by physiological and biochemical measures such as changes in heart rate, in the frequency at which a flickering light can be just detected, in electrical activity of the brain, and in blood and urine composition. Most of these techniques are not yet suitable for routine use in industry. To summarize, fatigue is an incredibly complex phenomenon for which there are no simple measures and on which any measurements, statements or observations made by investigators or by participants must be treated with the greatest caution.

This complexity is well exemplified in the problems of vigilance. In the typical vigilance or inspection situation the operator is not engaged in physical activity (i.e., he is sitting apparently relaxed), nothing happens for a long time, he is isolated from irrelevant interruptions, and he is looking or listening for something not easy to detect. Yet it turns out that performance in this situation is more susceptible to fall-off than in any physically active work even when this is extensive. This sort of working situation is becoming increasingly common in industry as the hard physical work is taken over by machines, but man is still needed because of the versatility and sensitivity of his sense organs. It has often been demonstrated that performance in such situations can fall off in less than half an hour. Typically, the probability of a missed signal (e.g., a just perceptible fault in an article being inspected) has increased after about a quarter of an hour, so also has the time taken to notice a signal. The lower the frequency of signals the less is the chance of detection. Non-optimal environmental conditions of lighting, noise, heat or cold reduce the level of efficiency. Short rest pauses can restore performance, so also can unexpected happenings either in the task or in the surroundings. Knowledge of results reduces the rate of decline. There are large individual differences; some subjects show no effects of this kind. The performance decrement is almost certainly psychological rather than physiological in origin, no changes in the sense organs can be detected, and performance can be improved abruptly by psychological manoeuvres such as informing the operator of missed signals, changing his task, and providing extraneous interruptions or incentives. In general, the more monotonous the task and the more difficult or subjective the judgement, the greater is the decline in performance. The situation arises commonly in industry when human operators are used to inspect the output of automatic processes. Matters can be improved by detailed attention to the environment (particularly to the lighting for visual inspection), by defining the task as accurately as possible, by frequent rest pauses, and by careful selection and training. Not much is known as yet about the criteria for selection except that, since it takes a long time to acquire these skills, operators who are not likely to change their job often should be used. There are a few self-evident requirements, such as good vision. Skill often depends on the long-term development of complex sets of visual criteria for " good " articles, and faulty ones then stand out as different from the expected. After long experience an inspector will be intuitively aware of the proportion of faults to expect under given conditions. This is an asset if he is right but can be a serious liability if he is wrong, since he will reject the number of items expected to be faulty, regardless of whether they are or not.

If an inspector is dealing with products in which other workers have been closely involved, there are bound to be problems of social relationships. The inspector is usually reporting directly to management, the production worker is not; the inspector is checking the worker, but the worker cannot check the inspector. Thus an inferior/superior relationship exists although the levels of skill required may well be in the reverse order. This is not helped by the above-mentioned inconsistencies in inspection, which will often be detected or at least suspected by the workers. In such situations working procedures and responsibilities for everyone concerned must be defined and agreed upon, and particular attention should be paid to channels of communication. Where inspection is critical, usually for reasons of safety, the probability of fault detection can be increased markedly by inspection and independent reinspection.

The general problems of safety and accidents are not usually approached in the traditional way by the ergonomist. For example, he will blame the situation rather than the operator even when the cause is clearly " human error ". Human errors are an important source of information about equipment design faults. Thus the attitude is often taken that, if the task has been properly designed from an ergonomics point of view, the solution of problems of errors and safety will have been arrived at almost as a by-product. The study of accidents as accidents is not usually fruitful for a number of reasons. Accidents are rare and very large samples are needed for statistical significance to be established. By the time this has been done there are usually so many variables that cause/effect relationships cannot be established. For example, in the city of Birmingham, England, motorists were asked to drive always on dipped headlights for one winter. During this five-month period the number of fatal accidents dropped to half that of the previous year. But even for such a long period in a very large city the actual figures were only 39 and 20 respectively. Moreover, it happened that this was a particularly severe winter and also that a great deal of general propaganda about safe driving had been disseminated. It is clearly impossible to draw any specific conclusions from such data. Accidents are usually literally accidental in their effects if not in their causes. Thus the same human error repeated on many occasions may have had no ill effects, and then, committed once more, it may result in a fatality. Furthermore, statistics can only be about reported accidents, and the tendency to have accidents may be unrelated to the tendency to report accidents. The effect of an individual's accident record on our opinion of his skill is also often unfair, in that exposure to risk is so varied.

In general, then, the ergonomist prefers to approach the problem from a " people-centred ", rather than an " incident-centred ", point of view. Various concepts of accident proneness have been developed, but some evidence suggests that the concept may not be as simple as it looks at first sight. The established statistical differences between individual accident rates with the same exposure may be due, in some situations at least, to the fact that each individual has spells when he might be more likely to have an accident, rather than that some individuals are consistently more likely to have accidents. This would account for the difficulties found in attempting to devise selection tests for the accident prone. This would seem superficially to be one obvious way of reducing accidents, but more practical possibilities are to reduce either the real risk or the apparent risk. Reduction of real risks by better equipment and task design is obviously valuable, but it may not change accident rates. For example, better acceleration and braking characteristics for motor-cars improve the design aspects, but unfortunately the driver often adjusts his tolerable margins of error accordingly so that there is no change in the accident rate. This is one illustration of the general point that accidents are more often related to attitude than to skill or equipment design. Most individuals have an unfortunate tendency to regard accident risk as applying to other people rather than to themselves. This is one reason why it is difficult to ensure that workers use the safety devices provided. Although this attitude may be susceptible to slight change by indoctrination and propaganda, the only real solution is to design in such a way that safety devices and procedures are an inherent and necessary part of machines and tasks, as well as being emphasized in instructions and training schemes—for example, a fuse box so designed that it is physically impossible to open it without switching off the power is obviously safer than one that merely has a notice on it saying " switch off power before opening box ". Another general procedure is to increase apparent risks rather than real risks. For example, it has been found by experience that inspectors of electrical equipment are safer when using metal torches rather than rubber torches, because the metal torch is obviously dangerous when manipulated within the equipment. Conversely, the particularly dangerous situations are the ones that appear to be relatively safe. Thus, in road design it is important that bends should appear to be at least as sharp as they really are. From the ergonomics point of view, then, the problem of safety is approached in just the same way as the problem of productivity. The abilities and limitations of the human operator are assessed in relation to the task, the work space, and the equipment, and all these are designed accordingly, as are also selection procedures and training schemes for the worker.

General Principles

1. The speedier and more complex and difficult the motor task, the more performance is likely to be affected by age, but for many high-level skills experience acts as a compensatory effect.

2. To reduce the effects of age on a particular task:

(a) reduce the required speed of work, particularly machine pacing;

(b) increase the lighting levels;

(c) improve the presentation of information so that less material needs to be carried in the memory; and

(d) design the work space so that strenuous movements or excessive postural changes are avoided.

3. In designing training schemes for older people, reduce the length of the learning steps.

4. Older people should not be expected to acquire habits that are in any sense a reversal of habits already established.

5. The prevention and reduction of fatigue is usually better acheived by frequent short rest pauses than by less frequent longer breaks.

6. To avoid muscular fatigue not more than about one-fifth of the maximum power of the muscle should be used, particularly in the case of static work.

7. For most industrial jobs, all kinds of fatigue are reduced by introducing more variety into the work and methods of working.

8. For many industrial tasks " rest " need only be a change of activity rather than a cessation of activity. Thus, if a team of operators regularly interchange tasks, fatigue effects should be smaller than if they all remain at their own particular task.

9. There is little correlation between subjective feelings of fatigue and performance effects. A worker may feel very tired but his performance may be unaffected, and conversely he may feel highly alert even after his performance has deteriorated.

10. Workers of different personality types react differently to fatigue: in confined situations some will become increasingly aggressive and overactive, while others will become increasingly withdrawn and under-active.

11. Fall-off in inspection performance can be reduced by:

(*a*) design of optimal environmental conditions;

(*b*) rest periods and frequent changes of task;

(*c*) increased perceptibility of the faults to be detected (e.g., improved contrast);

(*d*) knowledge of the results of performance; and

(*e*) improved selection and training methods.

12. Every inspector should be given detailed instructions—i.e., a precise indication of what constitutes a fault and what does not—to cover all possible contingencies.

13. Accidents and near-accidents are a useful source of evidence about bad task design.

14. All operators are subject to error and all tasks should be designed with this as an accepted constraint.

15. Accident-prone situations are more easily detected, understood and remedied than are accident-prone individuals.

16. Situations are particularly dangerous when the real risk is greater than the apparent risk.

Particular Cases

1. The lighting level required by an individual for a particular job is likely to double with every 13 years of age.

2. The willingness of older people to change jobs within a factory or company is usually confounded by status problems.

3. Older people are likely to be less tolerant of heat stress.

4. One of the commonest causes of fatigue is stressful posture. This can and should be remedied by better work design.

5. When a worker makes errors due to fatigue, he will often honestly believe that the machine rather than he caused the error.

6. Systematic studies of results of inspection often reveal startling inconsistencies. For example, three inspectors each inspected eight pieces of equipment; each of the pieces of equipment was rejected by at least one inspector and accepted by at least one inspector.

7. An experienced inspector will conserve his effort by concentrating on those aspects of his task that might be relevant to fault detection—e.g., if he is inspecting vacuum cleaners for defects and is asked the colour of the last vacuum cleaner inspected, he will not know the answer.

8. When there are repairers available to deal with the rejects directed to them by an inspector, the latter may easily come to feel responsible for

keeping the repairers busy and adjust his fault detection standards accordingly.

9. Unless intensive training is given, safety equipment such as guards and special clothing will be disregarded if the use of it entails reduced production or greater effort.

10. Urging people to be more careful is not usually a profitable exercise—the effects are only temporary.

———————

AGE, FATIGUE, VIGILANCE, AND ACCIDENTS

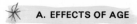

A. EFFECTS OF AGE

	EFFECT	CONSEQUENCES
RECEPTORS	General decrease in efficiency	For vision better lighting needed For hearing no consequences of importance to most work
CENTRAL PROCESSES	Decreased information handling capacity	Loss of 'thread' in complex routines Decreased versatility
PHYSICAL WORK PERFORMANCE CAPACITY	Some fall-off but not rapid until after 60 years	Increased importance of postural aspects of job design Heavy work can still be maintained unless speed is very high
COMPENSATORY EFFECTS	Increased experience Narrowing of interests	Greater range of situations already met Increased reliability

WHO 91265

B. CAUSE/EFFECT RELATIONSHIP IN FATIGUE

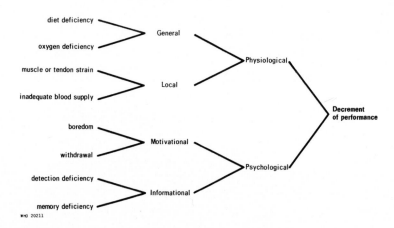

WHO 20211

C. INSPECTION DESIGN FACTORS

CONTEXT

Fault specifications
Social interactions

WORK SPACE

Lighting and contrast
Thermal comfort
Posture

OPERATOR

Selection and training
Rest pauses and job changes
Performance checks

WHO 91267

D. ACCIDENT PREVENTION

Design	Compensate for inherent human and hardware limitations
Programme	Provide adequate instructions and training
Maintenance	Maintain hardware performance and reliability
Operator performance	Ensure adequate motivation, prevent excessive fatigue, use adequate safety margins, maintain basic speed and accuracy requirements

WHO 91268

CHAPTER 11

Acquisition of Evidence about Individual Behaviour

All human beings, from their experience of themselves and other people, consider that they are in a position to generalize about human behaviour. These generalizations are often fallacious or too narrow, since one individual is likely to base his experience mainly on contact with a particular class of people in a particular kind of society. Again the individual is bound to acquire from his family, his education and his society in general a particular set of beliefs and prejudices that control his perception of other people's behaviour, not merely in value terms but in terms of what he genuinely does or does not notice or regard as important. This kind of evidence, founded only on ordinary experience, is clearly not good enough as a basis for a technology of behaviour. It is necessary to resort to more structured methods of acquiring information, of which there are basically three kinds: survey techniques, experimental techniques, and direct observation techniques.

Survey techniques depend on the principle of extracting information from people in situations that may be standardized but are not deliberately manipulated (if the situation is manipulated it becomes an experimental method rather than a survey method).

In *individual and group-testing methods* the subjects of study are exposed to a standardized situation. This may consist of questions on paper, as in intelligence tests, or a performance situation, as in fitness tests. The subjects' response to these standard situations is then codified by allocating a score based on performance. This score may be calculated from the number of correct answers given, the length of time taken to complete an exercise, or some measure of bodily function such as change in heart rate. Before such a score can be interpreted norms for the test must be provided. Norms are standards or levels expected from particular proportions of particular populations. Because of the range of individual differences an average is rarely adequate as a norm; some measure of where the individual fits in relation to the total spread of the population is needed and this is provided by percentiles. For example, the 60th percentile indicates that 60% of the population under consideration can be expected to score less than the stated norm and 40% will score more. For the reasons mentioned earlier in relation to anthropometry (see page 29), norms are usually stated either in terms of the 5th, 50th and 95th percentiles or in percentiles at 10% intervals. The purpose of a test is to determine in which category an individual belongs for a particular purpose, e.g., whether or not he is suitable for a particular job. It must be remembered that such a conclusion is never more than probabilistic. Thus a test score for an individual always indicates that that individual belongs to a set, and it may be known that a given proportion of the individuals in that set will achieve success at a given task. Of course, not even the most precise test can unerringly predict whether a particular individual will be one of those who will succeed or one of those who will not. It is for this reason that the psychologist has been defined as the only man who does not care what results he gets in a test—he is only too well aware of the fallibility of tests. However, they can be extremely valuable in categorizing and filtering populations for particular purposes. The accuracy of a test is measured in terms of two factors: reliability and validity. *Reliability* means consistency: if the same test is repeated on the same person the same result should be obtained. Such equivalence is unattainable, however, if only because, to be pedantic, it is never possible to repeat a test on the same person: the second time the test is given he cannot be the same person, since he has already done the test once and, in addition, is now older, as is also the tester. *Validity* means the accuracy with which the test score relates to the practical criterion such as job performance. Tests can be reliable and valid, or reliable without being valid, or at least potentially valid without being reliable. Physiological and anatomical tests are usually more reliable than psychological tests. For example, the measuring of a man's height before he joins a police force is a very reliable test but it may well not be valid since it is unlikely to predict the subject's efficiency as a policeman. On the other hand, a personality test of honesty and perseverance may be valid but is unlikely to be reliable. The criteria for a successful test are reliability, validity, availability of norms, and low cost, the last implying simplicity of equipment and ease of training for testers.

Controlled sampling methods are the same as test methods in that individuals are placed in standard situations and their responses are recorded and assessed. They are different from tests in that the purpose is to find out about or predict the behaviour of a particular group of the population. This is achieved by strictly random sampling or by stratified sampling. In a stratified sample the numbers within each layer or stratum of the sample correspond to the numbers of that stratum in the population being studied. For example, suppose the staff of a hospital consisted of 10 administrators, 20 doctors, 100 nurses, and 200 clerks, clearners and orderlies and they each had one vote in the selection of a colour scheme for a staff room, one might reasonably construct a sample of 1 administrator, 2 doctors, 10 nurses and 20 clerks, etc., and ask them how they proposed to vote. The accuracy of prediction from such a survey depends on the accuracy of answers and on the construction of the sample. For example, it might be better with the colour scheme problem to split the population into males, females, old and young and ask an appropriate sample constructed on these strata regardless of their profession. As far as accuracy is concerned it has been found that replies pertaining to the past or present situation are much more accurate than those pertaining to the future. Exactly how the questions are phrased or in what order they are asked can also affect the accuracy of answers. It is not so necessary to take such fine points into consideration in most physiological and anatomical surveys but, since in these cases the procedure is often more tedious for the subject, it is very important to adhere rigidly to the sampling rules whether the sample be random or stratified. The procedure for random samples requires just as close a statistical control as that for stratified ones, since extraneous variables may have unexpected effects. For example, in an attempt to acquire a sample of normal electroencephalographs it was found that the substitution of paid subjects for unpaid volunteers changed the " normal " EEG pattern.

Depth interview methods depend on the establishment of rapport between the interviewer and the person interviewed. The investigator uses his conversational skills to assess in detail the response of an individual to a particular situation. There is usually some structure to the interview although the subject will not be aware of it. This is a highly skilled technique requiring long experience and continuous practice. The investigator must maintain an entirely neutral attitude to the topic of investigation, yet at the same time he must see to it that a mutual interest is kept up, and adopt an entirely unbiased approach. This technique is particularly useful in the study of attitudes that are often very muddled and have not been previously sorted out by the subjects themselves.

Experimental techniques are based on the measurement of behaviour in designed, controlled, and manipulated situations. In principle, since behaviour is determined by many factors, we have a better chance of understanding it within a situation if we control the factors operating in that situation and measure the resulting behaviour when the known factors are deliberately manipulated. There are two kinds of experiment: single variate and multivariate. As the names imply, in the former everything is kept constant except for one variable, which is changed systematically, whereas in the latter a number of variables are changed but in an orderly fashion so that their effects can be separated by statistical analysis. Multivariate experiments are more economical but, particularly in relation to human behaviour, there are difficulties of interpretation due to non-linear interactions between factors and sequences of factors. For instance, if heat and noise conditions are varied systematically within one experiment to detect effects on performance, it is possible in theory to sort out the relative importance of each factor. The difficulty is that, taking a constant noise level, different levels of heat stress may add to, subtract from, or leave unchanged the effects of the noise. In addition, the whole experiment may be confounded by the level of performance achieved during the first exposure, which thereafter becomes the target level regardless of noise and heat.

In single variate experiments the most difficult problem is to provide a control, that is, a measure of another group of subjects or another condition that can be compared with the measure obtained when the factor being studied is changed. For example, if a new machine is being compared with an old one and the subject's performance is measured before and after the change, it is desirable to have a control group of subjects who continue to use the old machines in the same environment throughout the experiment, just to check that either no extraneous variables are affecting performance (such as change of work-flow) or, if there are such effects, they can be corrected for. Thus, before initiating any experiment it is best to discuss the design with a specialist in statistics and in human factors studies. An adviser skilled in statistics only may be misleading, since he may not be aware of the particular problems of behaviour effects and measurement.

Even when conditions have been properly balanced and controlled, only the reliability aspects have been dealt with and there remain the problems of validity. The fundamental difficulty about experiments is that a controlled situation is almost bound to be an artificial situation that will encourage artificial behaviour. It is not easy to extrapolate from such situations to normal behaviour in normal environments.

Direct observation methods are essentially ecological in that they involve studies of real behaviour in its normal context. The situations are natural because they are real, and the skill of the observer rests on his ability to perceive and record the behaviour without interfering with it. Thus the " unnoticed observational " method is by far the more common in this category. There are a few " noticed observational " methods, where the investigator deliberately uses his presence to change performance and to obtain relevant evidence from what happens, but these methods require a very high level of skill and experience. Usually the observer is attempting to be as discreet as possible but at the same time to observe acutely what the operator is actually doing. Because of the normally unsystematic acquisition of information by the human observer these techniques aim to ensure that he approaches the problem in a systematic and unbiased fashion and records his information in a standardized form so that comparison with other situations is possible. There are two kinds of techniques: methods study and skills analysis.

Method study techniques are used to record the sequential activities of an operator. They depend on the use of standardized symbols and recording techniques that require some preliminary training. *Process charts* use a simple set of five symbols originally standardized by the

American Society of Mechanical Engineers and called the ASME symbols. These assume that each working time unit can be allocated to either operation, inspection, transport, delay, or storage, for each of which there is a symbol. Skill is required in the use of these symbols, since their meaning changes according to the type of chart being prepared, for example, " storage " on a material-type process chart means that the work has been put to one side; on a two-handed process chart it may mean that the work is being held in the hand. For more detailed analysis various micro-motion charting systems are available. These use more elaborate symbolic systems, such as therbligs or predetermined-motion-time symbols. *Flow diagram methods* are used to record patterns of behaviour and movement. The activities of people, processes or materials can be described by these techniques using simple analogue models of physical space—e.g., line diagrams or string diagrams.

Automatic methods of data recording—e.g., cameras or videotape recorders—can be substituted for human observers. The best-developed techniques involve the use of the cinecamera: behaviour can be recorded in real time, at reduced speed, or at increased speed. All the data are accessible for review many times if necessary and the speed of review can be faster or slower than the speed of the observation. Film analysis is frequently used in association with micro-motion charts. The cinecamera can also be used to give controlled sampling units—e.g., in " *activity sampling* " techniques, where the times of observation are controlled in either a random or a regular fashion.

Skills analysis methods attempt to subdivide the activity into component skills. This is a much more involved problem than method study, since it is not possible to divide the skills into sequential time units. The skills overlap in time, and since skills are hierarchical it is difficult to know when to stop the analysis. It is always possible to go on breaking down the activity into smaller units. The level to which the analysis is taken depends on why it is being carried out. Broadly, a skills analysis might be required for the solution of a selection and training problem, or an interface design problem, or both. Usually the analysis is taken to the point where skills obtainable by selection and training respectively can be separated and the component skills can be regarded as suitable training units. For example, a skills analysis for a typist would stop at the point of specifying " reading skill ", since this will be obtained by selection. On the other hand, a skills analysis aimed at devising a training scheme for backward readers would take the analysis of the reading skill into much more detail. The result of a skills analysis is called a *job specification*. Jobs can always be separated into perceptual and motor skills but division beyond this will vary with the task and the objective of the study. There have been various attempts to classify postural changes—e.g., to discover strategies of lifting and carrying—and perceptual activities—e.g., to try to determine the basis of decisions on the part of a process controller—but these are still not standardized to the extent that they can be taught easily to others. In summary, the techniques of skills analysis are much less formal than those of method study and require correspondingly greater creative insight on the part of the investigator.

ACQUISITION OF EVIDENCE
ABOUT INDIVIDUAL BEHAVIOUR

A. PROCEDURES FOR ACQUIRING EVIDENCE

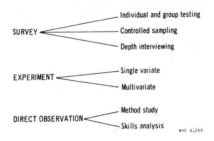

SURVEY
- Individual and group testing
- Controlled sampling
- Depth interviewing

EXPERIMENT
- Single variate
- Multivariate

DIRECT OBSERVATION
- Method study
- Skills analysis

B. SOURCES OF EVIDENCE ABOUT INDIVIDUALS

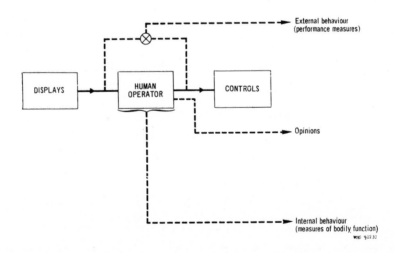

C. DETECTION OF ELECTRICAL VARIABLES

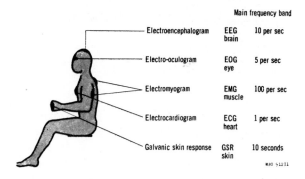

		Main frequency band
Electroencephalogram	EEG brain	10 per sec
Electro-oculogram	EOG eye	5 per sec
Electromyogram	EMG muscle	100 per sec
Electrocardiogram	ECG heart	1 per sec
Galvanic skin response	GSR skin	10 seconds

WHO 91271

D. PROCESS CHART SYMBOLS

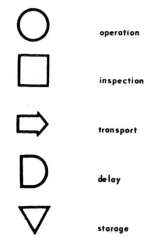

○   **operation**

☐   **inspection**

⇨   **transport**

D   **delay**

▽   **storage**

WHO 91272

The objectives of the process chart are to provide comparable records, to enforce systematic study, and to provide evidence of the need to introduce changes.
The above symbols are used to classify activities in the form of various kinds of process chart. They can be used for: flow process charts; man-type process charts; material-type process charts; and two-handed process charts.

CHAPTER 12

Acquisition of Evidence about System Behaviour

The methods of studying large and small groups parallel the methods of studying individuals. It is possible to use survey techniques, experimental techniques and direct observation techniques. In all cases the techniques are less well refined than for individuals. In using survey techniques, for example, the problems of classification and criteria have not yet been solved satisfactorily. There is a subject called *organization theory*, in which attempts are being made to determine the characteristics and parameters of organizations. In one such group of studies the accepted dimensions of organization—specialization, standardization, formalization, centralization, configuration and flexibility—have been studied within a wide range of industrial companies. Using techniques of principal component analysis four underlying factors have been identified and named as follows: structuring of activities, line control, concentration of authority, and strength of supporting services. It is then possible to go on from this analysis and try to define the context of the organization that is likely to result in a particular configuration of these factors. For instance, how the configuration is effected by origin and history, size, ownership and control, objectives, location, and branch of technology. The general hypothesis is that the behaviour of organizations is no more accidental than the behaviour of individuals and can be predicted, at least statistically, from a knowledge of latent characteristics and past history.

The criteria problem is usually regarded as coming within the province of accountants, who prepare balance sheets, profit measures, returns on capital and so on. It is difficult to draw inferences from such measures and relate them to the factors that affect individual and organizational efficiency, because these measures are at once too global and too short term: too global in that the number of internal and external factors that might affect the data is so large that separation is impossible; too short term in that individuals and organizations have expected lives of 50 years or more, so that returns based on one-year performances can be very misleading in relation to the long-term trends. It was fashionable in the years just after the Second World War to conduct inter-firm and inter-country comparisons of productivity, but again this raises large issues of comparability, given different conditions and objectives.

Organizational experiments have, for obvious reasons, been concentrated mainly on small groups. The performance measure in such cases is usually efficiency in problem-solving with controlled variables such as type of leadership and type of communication network. On the one hand, if the groups are too small the performance of individuals will dominate the organizational variables; on the other, if the groups are too large the situation becomes unwieldy and expensive.

Direct observation techniques range from the highly subjective methods of the management consultant to controlled sampling methods, such as activity sampling mentioned in the previous chapter (see page 114). It is true that an experienced consultant can walk through a factory and get an impression of the tempo and effectiveness of the organization, on the basis of which he can suggest, for example, that there is potential for a productivity increase of, say, 25%. Given a wide experience of that kind of factory making that kind of product he may well be accurate to within about 5%. It might be argued that such a procedure is too subjective to warrant mention in a technological context but, in fact, every technologist—whether he is an ergonomist, a civil engineer or anything else—is sometimes faced with the need to make such decisions. However, most ergonomists do feel more comfortable if they have the possibility of basing predictions and recommendations on more systematic studies. Relevant evidence can often be obtained by tracing communication networks—who tells whom about what, and how he does it. Such studies are often tedious and expensive but they provide great insight, since even the senior members of organizations are not usually fully aware of how the communication and control systems function. Attempts to draw *formal organizational charts* can be a useful discipline but they have at least two limitations. Firstly, there is often a considerable difference between the formal and the actual hierarchies of decision-making and influence. Secondly, relationships between a group of people are too complex to be at all adequately represented by straight lines. Because of overlapping responsibilities, *Venn diagrams*, which represent overlapping sets, are a more accurate representation than line diagrams. It is particularly difficult to represent the distinction between line reponsibility and functional responsibility for the growing numbers in the service sections of companies. For example, a factory medical officer will usually have line responsibility vis-à-vis the factory manager and functional responsibility vis-à-vis the group medical officer. Because of the complexity of organizations, sampling techniques are a particularly useful tool. It is possible, with activity sampling over a relatively short period, to obtain a reasonable estimate of the total activity pattern within a factory—for example, the frequency of use of machines, corridors, telephones, etc.

Critical incident methods are a useful tool for the determination of stringent task demands. Data on critical incidents such as accidents, near accidents or emergencies are collected and studied so that the various categories with a common factor can be ascertained. These common factors themselves reveal critical times or kinds of occasion that are associated with special danger. For example, if all the press operators who have lost a thumb in an accident say that this happened when a particular kind of knife was used on a particular kind of material, then that knife and material are worthy of special study. Alternatively, if the accidents in question commonly occurred when the operators were engaged in conversation then preventive measures can be applied, such as training operators not to talk while cutting.

Task description specifies the activities carried out by human operators within a system or organization. It consists of two parts: the range of alternative strategies (courses of action) available to the operator and the information needed to choose effectively between those courses of action. A task description can be arrived at by *task synthesis*, which is the logical construction of the task in terms of objectives of the organization and the characteristics and skills of the available human operators. Alternatively, it may be arrived at by *task analysis*, which is based on the study of what individuals actually do as elements within systems. This is increasingly difficult with increasing levels of skill, partly because of greater complexity and also because often the more skilled an individual is the less aware he is of his own activities and strategies. In principle a task description can be arrived at by all the methods already mentioned, from surveys to depth interviews. It is an activity that runs parallel to skills analysis. The difference is that in task analysis the objective is to find out what the operator does as a system component, while in skills analysis the objective is to find out how he does it. The former is expressed in system, organizational or hardware terminology, and the latter in behavioural or biological terminology. A job specification resulting from a skills analysis is in principle convertible to a task description and *vice versa* by a change in terminology and an allocation of tasks or jobs. Thus a task may require more than one individual or an individual may do more than one task. By contrast, a job done by an individual may be composed of more than one or constitute less than one task. For example, a receptionist has the job of looking after customers but she may also do other jobs such as answering telephone calls and dealing with correspondence. These jobs together make up the total task of reception. On the other hand, a nurse may carry out the task of reception as part of her job in a surgery.

It is important to make the distinction between jobs and tasks, as well as the more general one between individual activities and organizational activities, if only because the activities of an individual cannot be understood without some knowledge of their place in the activities of an organization. Similarly, the ergonomist cannot make effective and adaptive proposals without considering the influence of the larger system on the part he is directly concerned with. For example, it is not possible to design a training scheme simply in terms of the job and the available people.

Whether the training scheme will survive or not is partly, if not mainly, a function of the attitude of higher levels of management and the way the scheme is built into the total organization. In more general terms the successful ergonomist must be something of a politician, a manager and an economist, as well as a behavioural technologist.

There may, at times, be some conflict between the needs and objectives of operators and those of systems. For example, it is possible for one operator to be too efficient, in that he produces so much from his job that he unbalances the activities of the team. More frequently the ergonomist has a problem in that too much of his knowledge and expertise relates to the individual and he is not sufficiently aware of, or expert in, the larger problems. For example, the performance of an individual at a task measured in terms of effort, speed and errors may be difficult to translate into the more general criteria of cost and system effectiveness. This question of *systems-relevant criteria* frequently arises, perhaps because the ergonomist has too dominant an inheritance of laboratory-based education and research. It will be argued in Chapter 14 that it is part of the responsibility of the ergonomist to translate experimental findings into operational evidence.

ACQUISITION OF EVIDENCE
ABOUT SYSTEM BEHAVIOUR

A. FACTORS CONFOUNDING FACTORY STUDIES

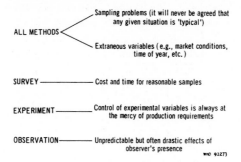

ALL METHODS
- Sampling problems (it will never be agreed that any given situation is 'typical')
- Extraneous variables (e.g., market conditions, time of year, etc.)

SURVEY ——————— Cost and time for reasonable samples

EXPERIMENT ———— Control of experimental variables is always at the mercy of production requirements

OBSERVATION ———— Unpredictable but often drastic effects of observer's presence

WHO 91273

B. SOURCES OF EVIDENCE ABOUT FACTORY STATUS

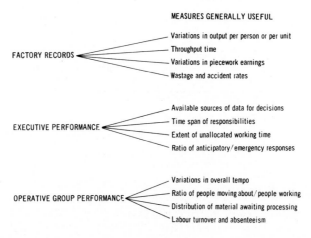

MEASURES GENERALLY USEFUL

FACTORY RECORDS
- Variations in output per person or per unit
- Throughput time
- Variations in piecework earnings
- Wastage and accident rates

EXECUTIVE PERFORMANCE
- Available sources of data for decisions
- Time span of responsibilities
- Extent of unallocated working time
- Ratio of anticipatory/emergency responses

OPERATIVE GROUP PERFORMANCE
- Variations in overall tempo
- Ratio of people moving about/people working
- Distribution of material awaiting processing
- Labour turnover and absenteeism

C. TASK DESCRIPTION

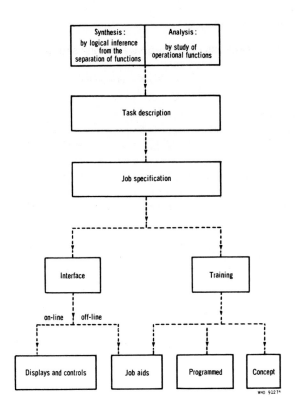

Synthesis: by logical inference from the separation of functions	**Analysis:** by study of operational functions	

Task description

Job specification

Interface — on-line / off-line

Training

Displays and controls | **Job aids** | **Programmed** | **Concept**

WHO 9127⁴

D. METHODS OF TASK ANALYSIS

Method	Principle	Recording medium
Functional indentification	Separate different activities carried out by operator and their interaction with process	Block diagram or path/node diagram
Dynamic interaction	Study operator responses, their causes and the time relationships	System state diagram or real time diagram
Robot simulator	Define how operator's job would have to be done by a robot	Chart or computer programme
Responsibility definition	By discussion and observation identify responsibilities and appropriate strategies	Job description chart

WHO 91276

CHAPTER 13

The Design of Work

The range of sciences and technologies that have some relevance to the design of any job is so large that it is difficult to find a single person with all the required skills. It is therefore desirable to construct an overall conceptual framework into which the various facets of work design can be fitted. This enables each specialist to perceive his contribution within the total context and also makes it easier to detect omissions in particular programmes. The exigencies of any real situation will probably be such that it is not possible to follow a logical sequence in practice, but this does not detract from the value of a systematic model as a referent and guide.

The only factor common to all work design problems is the human operator. Given that the objectives of the system have set the requirements of the task the problems connected with human operators can be separated into job design, work design and environmental design.

Job design includes selection, training, instructions and incentives. *Selection and training* problems are inseparable, since the required skills as described in the job specification can, in principle, be obtained either by selecting only operators who have these skills or by imparting these skills to operators by training. In practice, some skills are demanded as a condition of selection and others are acquired by training and experience. The selection criteria for operatives are often simply that the recruit must have no obvious physical deficiencies and must have reached a particular educational and age level. Such criteria can be assessed by a simple interview. At the other extreme, selection may involve elaborate measures of physical and sensory abilities, motor co-ordination, mental abilities and capacities, and personality characteristics. Standards can be laid down both in terms of level required and in respect of how this level should be measured by the use of specialized tests. The design of selection methods is the province of the occupational psychologist. So also is the related problem of training design. The method may vary from requiring the trainee to spend some time observing an already skilled operative to letting him undergo a long period of off-line training. The decision on whether training should be on-line or off-line can only be made in relation to particular jobs. Off-line training has considerable advantages in that the objective is clear and not confused by the demands of the production process, exposure to all the required skills can be balanced, and it is often less dangerous and less expensive, since the trainee is not producing spoiled or below average products. The design of all training schemes is based on knowledge of the advantages of whole and part methods of learning. Whole methods have the obvious attraction that the job eventually has to be done as a whole anyway, and even if a trainee acquires considerable ability in all the parts separately he will still have to learn to integrate his skills. On the other hand, part methods have the tremendous advantage that knowledge of the results of work and effort can be provided more frequently and more accurately. It is well established that " knowledge of results " is the key to improving performance and also to providing the incentive to learn. It is often possible to use " progressive part " methods of training in which the parts are learnt separately, with efficient knowledge of results, but are combined progressively rather than all at once at the end of part learning. Training in the parts of a task usually requires the use of training devices. The design of these is a specialized art, the criteria used being fidelity (that is, how closely the task on a device represents the real task), transfer (that is, how readily learning on the device transfers to the real task), and cost. The objectives of a systematic training scheme may be to increase ability, to reduce learning time and cost, to decrease the required entry standards or, most often, some weighted combination of all three. It is important that these general objectives should be discussed and agreed upon as a first step in the design of any training scheme. The more specific objectives, based on the job specification, the needs of the company, and the level of available trainees, can then follow.

The *provision of instructions* as part of the job design is a separate problem in that these may change more frequently than the required skills. If the operative is to be efficient he must know what he is trying to do, the particular problems he is likely to meet, the standards of speed and quality expected of him, and so on. Instructions that are unlikely to change with time, such as directives on how to set up and maintain a machine and the design of training exercises with time and quality standards, should be incorporated in instruction manuals.

The operator must be adequately motivated. The *provision of incentives* is often no more complicated than the agreement to pay at a given rate per unit of work produced. This may not be sufficient unless other aspects of job design, such as machine maintenance, supply of materials and the provision of a reasonable working environment, are carried out with efficiency. Longer term aspects such as preservation of health and prospects of promotion must also be considered. External factors such as travelling conditions, home environment, diet, and general family circumstances will also affect the working effort. Even for the least privileged workers the group interactions and atmosphere inside the factory and the general satisfactoriness of the community outside the factory will have a considerable influence on work.

Work space design interacts with the design of the job. The physical dimensions of the work space and the associated postural aspects of the work are relatively straightforward matters. Nevertheless they need to be considered systematically. The dimensions are dependent not only on body size, but also on the kind and direction of forces to be exerted and the facilities for providing the operator with information and for receiving his decisions. These are respectively the power and information aspects of the interface. Details for a particular job should be available from the task description and job specification. A job may involve setting up, on-line operation, or maintenance, or some combination of the three. These components are usually best considered separately since they set different problems in terms of both type of operator and the hardware interface design variables. The information provided for a programmed operator, who does his job by following detailed rules and instructions, will be different from that provided for a concept operator, who decides what to do in particular circumstances on the basis of a comprehensive under-standing of how the system functions. If the interface is at all complicated the problem of grouping displays or controls, or both, will arise. Grouping is usually done in one of three ways:

(*a*) by sequence, for operations such as setting up or monitoring, in which the operator is expected to perceive or to do certain things in a fixed order; the displays or controls can be placed in this order, from left to right or from up to down, or both—e.g., the controls for activating all the electric motors, pumps and gear trains on a large machine tool;

(*b*) by priority, if there are certain interface elements that are much more important or more frequently used than others—e.g., the speedo-meter on a car or the space-bar on a typewriter;

(*c*) by function, if there are several distinct activities going on within the system—e.g., in a power station the boiler displays and controls might be separated from the generator displays and controls.

The design of interfaces cannot be reduced to a fixed sequence of decisions that can be followed in every case. All the variables interact with each other and it is usually best to start by fixing the ones that are likely to have the greatest effect on efficiency. For example, on an automatic lathe the " setting-up " interface elements are likely to be given a higher priority than others, but on a manually operated machine the " on-line " elements will be dealt with first.

Environmental design is a global term covering three separate aspects: the physical surroundings, the working surroundings, and the surrounding people. In some jobs—e.g., driving—the *ambient environment* may have to be considered in relation to the individual, but more often it is a general problem affecting a group of operators. Even in this latter case, the problem of lighting is separate, and it is usually best to provide some supplementary lighting for every job. Consideration of heat and cold will also depend on the heat generated by the operator himself—that is, his physical work rate and also the heat output from his machine or associated with his task. For example, a furnace operator obviously has special problems, and so also does an overhead crane driver if he is moving over molten materials. Provided that they do not approach levels at which there is a risk of damage, noise and vibration effects will be much less important if they are generated by the job than if they come from outside. (It may be noted, for example, that a man riding a motor bicycle is not disturbed by these effects, but the people he passes may well be).

The *working environment* contains many other variables of this kind. A well-designed job in a well-designed work space will not be satisfactory unless the supply of working materials and the hours of work have also been properly determined. Again, these problems apply to a group of jobs rather than to one in particular, but their effect on the individual operator is often critical.

Such features of the *social environment* as communications, responsibilities and working groups also present problems. These can be dealt with partly under the heading of job design by providing clear job definitions and instructions, but the adaptability of the operator and the everchanging requirements of his job lead to considerable dependence on day-to-day contact with other people. These interactions are essential not only for the maintenance of morale but also to ensure that the system stays in equilibrium. Every production planning system and team control depends, for fine adjustments, on the co-operative behaviour of people doing different jobs, and this co-operation can only be based on communication.

There is, of course, an inherent contradiction between the narrow activities of, say, the heat stress specialist or the dial designer and the need to consider the way in which any specialized design problem depends on every other design aspect of the job. This can only be overcome by awareness on the part of the specialists of their role in the total scheme and by the activities of general ergonomists. These latter will consult their specialist colleagues, but they will have to carry the overall responsibility for the job design. At present, this is often in fact left to the engineer or manager, and will continue to be until there are enough ergonomists available to undertake the necessary tasks.

WORK DESIGN

A. ASPECTS OF WORK DESIGN

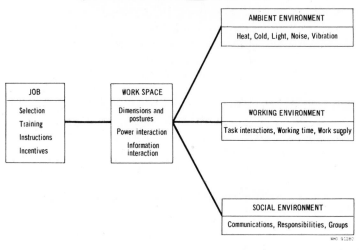

B. TRAINING PROCEDURE

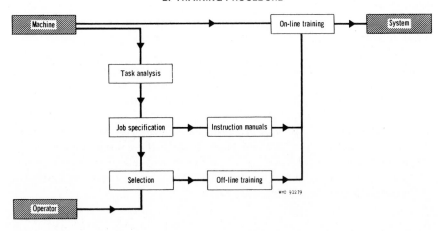

C. TRAINING METHODS

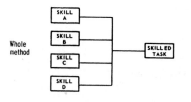

Whole
method

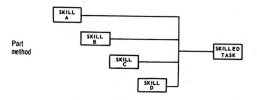

Part
method

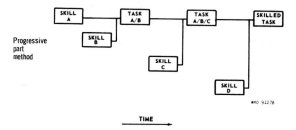

Progressive
part
method

WHO 91278

TIME ⟶

D. DESIGN OF INTERFACES

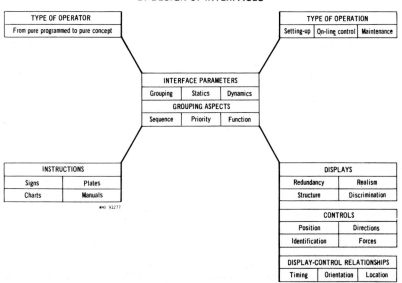

TYPE OF OPERATOR
From pure programmed to pure concept

TYPE OF OPERATION		
Setting-up	On-line control	Maintenance

INTERFACE PARAMETERS		
Grouping	Statics	Dynamics
GROUPING ASPECTS		
Sequence	Priority	Function

INSTRUCTIONS	
Signs	Plates
Charts	Manuals

WHO 91277

DISPLAYS	
Redundancy	Realism
Structure	Discrimination

CONTROLS	
Position	Directions
Identification	Forces

DISPLAY-CONTROL RELATIONSHIPS		
Timing	Orientation	Location

CHAPTER 14

Assessment, Presentation, and Interpretation of Evidence

As a technologist, the ergonomist has two problems: one of assessment, in which he determines what should be done, and one of implementation, in which he has responsibility for ensuring that something is done. The two are interdependent and equally essential. It is of no value to know what needs changing unless some change results and correspondingly it is of no value to take action without the guidance of technical expertise. The two problems overlap when the question of what is desirable is replaced by the question of what is feasible.

Within the area of assessment of evidence there is a distinction between statistically significant differences and differences that may be of practical importance. Taking the simplest problem first, statistical significance depends on the theory of sampling. Sampling problems are present in every scientific investigation but they are often neglected, with justification, by natural scientists and engineers. If the variables to be studied are sufficiently small in number so that isolation is possible, and if measurement is reliable in the sense of consistency, then a difference between readings, or at least between averages, can be regarded as significant without further exploration. This is rarely, if ever, sufficient in the biological and behavioural sciences. The large and inextricable combination of variables affecting even the simplest laboratory situation is such that there will be some unreliability. That is to say, when a measure is repeated under apparently identical conditions the results obtained may well be different. Continued repetition of the same measure under identical or similar conditions will result in a distribution of readings that is not adequately described by merely stating the average. Fortunately the number of factors or variables is usually so large and their effects are so random that the end result of any set of biological measurements is a normal distribution. This is simple and convenient in that the average adequately represents the central tendency of the data and in fact the average, or mean, is the same as the mode (the most common reading) and the median (the middle reading when all readings are arranged in rank order). In addition, the variance of the distribution is a well-defined measure of the scatter and range of the readings. Thus, for most statistical purposes, the mean and variance of a set of biological data are all that is required. It is usually too expensive or too tedious to continue taking readings indefinitely and so only a sample of all the possible values is obtained.

Inferences from such data usually depend on its being possible to determine, when something has been measured under two different conditions, whether or not the difference between the averages is accidental because of the sample limitations or whether it does represent some real difference between the situations in which the two sets of data were obtained. Confidence in stating that there is a real difference will obviously increase with the increased size of the difference between the averages, the increased size of the samples, and the decreased size of the variances. (The greater the scatter and range of the distributions the more likely it is that a difference of averages will occur accidentally.) The formalized mathematical logic of this problem, including a numerical statement of " confidence " in the form of a probability, is typical of statistical sampling methods. This particular one is dealt with by applying the so-called t-test. For example, in designing trucks for India we may wish to know whether drivers in Bombay are bigger or smaller than drivers in Calcutta. We can measure the height of drivers in the two cities but, of course, we do not want to measure too many for reasons of economy even though we know that there will be an overlap in the distributions. That is to say, whatever the answer to the question, some drivers in Bombay will be taller than some drivers in Calcutta and *vice versa*. However, for a quite small sample of drivers it will be possible, on the basis of this statistical calculation, to say at a given level of probability whether or not the obtained difference in average height is " significant "—or, in other words, whether or not there is a real difference.

Another, more advanced, statistical test depends on taking the total variance of a distribution and splitting it into parts depending on the factors that can be assumed to have caused those parts of the total scatter. It is then possible to compare the relative sizes of these parts of the total variance and assess which, if any, are the key factors causing the variation. This very commonly used technique is called *analysis of variance*. For example, in a heat stress experiment, variation of, say, performance measurement will depend partly on the fact that different individuals have been measured and partly on the fact that measures have been taken at different temperatures. Given careful control of the way in which subjects and temperatures are varied it is possible to calculate how much of the performance variation is due to different subjects and how much of it is due to different temperatures. Given this and some measure of all the other factors, it is possible to assess whether the individual differences are " significant " and whether the temperature changes are " significant "— that is, whether such performance changes could reasonably have happened by chance or not. These are *parametric statistical methods*, which assume that the measures taken are of the same kind as those obtained in the natural sciences—that is, 10 is greater than 5 and is also twice as great as 5. In the behavioural sciences the problem can become very much more complicated in that this need not be true. For example, in an opinion survey of noise we might set up a scale in which 4 represents " intolerable " and 2 represents " objectionable ". For condition A we might get an average rating of 4 and for condition B we might get an average rating of 2, but we cannot assume that A is twice as noisy as B. For this type of situation there are other, so-called *non-parametric statistical methods*. Different tests make different assumptions about the scales of measurement used and it is necessary to select a test in which the assumptions made about the measurements are valid. In general, statistical methods are a valuable aid to the construction of inferences from data but, of course, they are not a substitute for care and rigour in the acquisition of data.

It will be clear also that although statistically controlled data acquisition and inference can ensure the economy and reliability of studies there is still the separate question of the practical importance of significant differences and other findings. For example, to return to our trucks for India problem, Bombay drivers may be taller than Calcutta drivers at a very high level of significance but the difference in averages might be only one centimetre, which is of no consequence in cab design.

Unfortunately there is no methodology for establishing importance that corresponds in any sense to the statistical methodology of significance. The basic objectives of ergonomics are improvement of productivity and protection of health. In both cases the contribution is usually indirect. Direct damage to health comes within the province of industrial hygiene and toxicology rather than ergonomics. The ergonomic contribution is usually long term and therefore difficult to demonstrate, and it is correspondingly difficult to convince workers and managers of its importance. There remain large areas of ignorance about the effects on health of such factors as bad posture, non-optimal environmental conditions and information overloads. Even when evidence is available it is not necessarily used. For example, the relatively poor life expectation of dentists is almost certainly connected with working conditions such as posture, but only recently and on a small scale have positive steps been taken to improve the situation. Similarly, the worker exposed to high noise levels will often not regard the evidence that he may suffer some hearing loss after 30 years as sufficient to offset the inconvenience of wearing ear defenders. The increasing frequency of cardiovascular disorders in excecutives has not persuaded very many of them to change their way of life.

The same problems obtain in relation to working efficiency. There are examples of ergonomic studies leading to increases in output and productivity, but more frequently the savings are indirect in the form of the same output for decreased effort, a reduction of training time, an improved quality of work, and a declining number of accidents. There are also long-term effects deriving from the improved reputation of the company as an employer of labour, which contributes to a reduced labour turnover and easier recruitment of new workers. It must be acknowledged that it is more than usually difficult to obtain reliable and valid evidence on all these effects, but this is not sufficient reason for assuming either that they are unimportant or that they deserve a lower priority than improvements in easily measured variables such as productivity.

The intermediate area between efficiency and health is described by terms such as " comfort ". These are extremely difficult to define precisely and are correspondingly difficult to use in any scientific investigation. Even the objectives of such studies are not free from ambiguity, since an increase in comfort is not necessarily a satisfactory criterion for anyone. Most managers do not regard it as a good reason for spending money and many workers look upon some discomfort as a justification for higher pay or for a higher status among their fellow workers. All this supports the point of view that ergonomics is best regarded as being concerned with health (in the sense of total well-being) and efficiency (in the sense of optimal use of human labour). It must be stressed that in an evaluation of the contribution of ergonomics to these fields a comprehensive approach should be adopted; simple measures such as day-to-day productivity and absence of disease should not be the sole criteria.

This situation makes the presentation of evidence one of the major aspects of ergonomics. There have been too many occasions on which available ergonomics data have been ignored. There have also been too many ergonomists who have avoided their responsibility by pointing out that the work was done and the recommendations were made but these were not taken into account. It is part of the responsibility of any technologist, when he has a valid case, to present it in such a way that it will not be ignored by reasonable decision-makers. The common failings are an excessive use of specialist jargon and an unjustified assumption that the policy-makers will be as interested in ergonomics studies as the investigator himself is. Both these defects can be mitigated by bearing in mind that the policy-makers are normally not interested in the details of how a study was done but only in what it reveals; they are probably more concerned with the costs, benefits and other consequences of the recommendations than with the conceptual reasoning that led to the recommendations. These are difficult lessons to learn for the dedicated scientist or technologist, but such considerations dominate his practical success. It has often been pointed out that the real difference between successful and unsuccessful consultants is not so much in the quality of their technical studies but in the care and expertise devoted to the presentation of results. Looked at more technically, this is a communication problem: for the efficiency of any communication system the properties of the destination are just as important as the properties of the source. How the technologist encodes his information must be dominated by his estimate of how his customer will operate as a decoder of that information. To put the same point in yet other terms, the ergonomist must use systems-relevant criteria rather than scientific criteria. For example, take the question of accessibility for the driving position of a local delivery vehicle. Differences in ease of access cannot be used as direct criteria governing the choice between two alternative designs if the easier model costs more. It then becomes a question of cost and availability of drivers, proportion of time spent going in and out of the vehicle, the attitudes of the vehicle owners as distinct from those of the drivers, and so on. All these factors will be assessed by the efficient systems ergonomist before he presents his report, unless he has been formally restricted by his terms of reference.

The cost in terms of time and effort spent in drawing up the report may constitute a considerable proportion of the total cost of the project. There is nothing to be gained by economies in quality of stationery, script, diagrams and layout, since the objective is to make interpretation easier for the user and also to encourage a proper respect for the work done. It is not unusual to find that a study is condemned or ignored because of the quality of the report rather than the quality of the work.

In the implementation phase a major role that the ergonomist must play is based on his more acute awareness of the probabilistic evidence on which his recommendations were based. By definition he could well be wrong; although, as an expert, he is less likely to be wrong than the layman, this should not be allowed to obscure the fact that he will inevitably make mistakes. It follows that all implementation procedures must be flexible and adaptable and that there must be built-in validation techniques. That is to say, it must be possible to check the extent to which the predicted effects have followed the changes. This may be effected at any level, from occasional visual checks to the systematic acquisition of evidence under controlled conditions, but it is always desirable at some level to ensure that ergonomics as a subject continues to advance, as well as to check the particular case.

Another role for the ergonomist, arising from his experience of change and his expertise in dealing with people, is in the general psychology of innovation. In any large system, such as a factory, the prospect of change will have perturbations that will affect the whole organization and every person concerned, however indirectly. The well-established human relations techniques of communication and consultation are essential to overcome resistance to change, which manifests itself even when the change appears to be clearly for the better. The fact that a particular level of skill and effort was originally required simply because of bad work design does not alter the fact that the workers who have acquired the skills in question will naturally take some pride in them and will resist their elimination. Likely improvements in financial rewards, status, effort, health and so on must be carefully communicated to those concerned. One common mistake is to assume that, because a process or a machine has been well designed ergonomically, it can simply be substituted for an old one and left to the workers to operate. No matter how well designed the new system is there will always be some need for induction and retraining. Many experiments involving the introduction of a machine with innovations have failed, not because the changes were wrong but because insufficient care was taken in the method of introduction.

It is usually best to assume that every person involved in a change is going to be difficult, obstinate, conservative and even stupid at first. Paradoxically one can also assume that if enough time, patience and effort are devoted to telling everybody exactly what is going to happen almost everybody will eventually be reasonable, and if the innovation is planned on these assumptions its chances of success increase enormously.

ASSESSMENT, PRESENTATION, AND INTERPRETATION OF EVIDENCE

A. SIGNIFICANCE OF DATA

STATISTICAL	PRACTICAL
Probability that observed effect has occurred by chance	Probability that observed effect is important in some real situation

Magnitude of effect

Sample sizes

Identification of variables
  in situation design

Power of statistical method

Health
  Exposure to disease risk
  Exposure to accident risk
  Personal well-being

Efficiency
  Cost effectiveness
  Effort
  Productivity

WHO 91281

B. PRESENTATION OF EVIDENCE

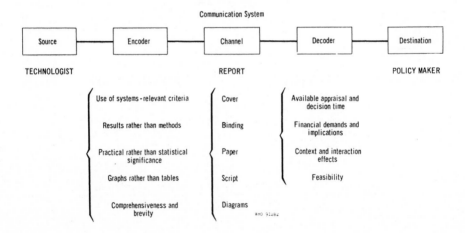

Communication System

| Source | Encoder | Channel | Decoder | Destination |

TECHNOLOGIST   REPORT   POLICY MAKER

Use of systems-relevant criteria

Results rather than methods

Practical rather than statistical
  significance

Graphs rather than tables

Comprehensiveness and
  brevity

Cover

Binding

Paper

Script

Diagrams

WHO 91282

Available appraisal and
  decision time

Financial demands and
  implications

Context and interaction
  effects

Feasibility

C. MEANS OF OVERCOMING RESISTANCE TO INNOVATION

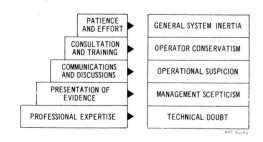

PATIENCE AND EFFORT	GENERAL SYSTEM INERTIA
CONSULTATION AND TRAINING	OPERATOR CONSERVATISM
COMMUNICATIONS AND DISCUSSIONS	OPERATIONAL SUSPICION
PRESENTATION OF EVIDENCE	MANAGEMENT SCEPTICISM
PROFESSIONAL EXPERTISE	TECHNICAL DOUBT

WHO 91465

D. COMPARISON OF ERGONOMICS WITH OTHER BEHAVIOURAL TECHNOLOGIES

	ERGONOMICS	WORK STUDY	OPERATIONAL RESEARCH	CYBERNETICS
ORIGIN	Experimental human sciences	Shop-floor production problems	Mathematics and economics	Control system theory
FAVOURED PROCEDURE	Inferences from theory and experimental evidence	Generalizations from experience and observation	Construction of models	Simulation
VALIDATION	Statistically controlled experimentation	Before/after comparisons	Check of model predictions against real occurrences	Comparison of simulated behaviour with real behaviour
SPECIAL INTERESTS	Human operator based systems	Human activity	Total systems	Biological/engineering analogies
SPECIAL TECHNIQUES	Acquisition of behavioural evidence	Estimation by human observers	Various standard mathematical models	Comparisons of control systems
ADVANTAGES AND LIMITATIONS	Systematic and slow	Fast and superficial	Elegant but often oversimplified	Theoretically interesting but difficult to apply

WHO 91284

CHAPTER 15

Retrospect and Prospect

Ergonomics is not a new subject in the sense that work of this kind was not done until the subject was invented. Ever since the days when man first ventured out of those limited regions of the world where he could be comfortable all the year round without clothing and other artificial aids to environmental adjustment, or the even earlier days when the first tools were invented, there have been individuals sensitive to their own and other people's problems of adaptability. These individuals have approached the resulting design questions with intelligence and eventually with accumulated experience. It is only during the present century that this empirical system has shown signs of inadequacy. Several features of development accelerated the need for a more scientific approach. One such feature consists in the increased specialization of individual skills, which results in the design of machines, tasks and environments by people with little practical experience of their use and little contact with those who have to use them. If an agricultural implement is designed by a local blacksmith who is himself a part-time farmer living in a village in daily contact with other farmers, the design errors in the implement from the user's point of view will be removed quickly. On the other hand, when a draughtsman is working in a factory on the design of a tank for soldiers he has never met and for conditions he has never experienced he is unlikely to produce a completely satisfactory design. Moreover, since wars usually last only a few years, there is no time available for very highly specialized training or for the processes of communication and trial and error gradually to take effect. In any weapon design the key problems are to ensure correct functioning as quickly as possible and not to leave too much to the discretion of the user. It was, in fact, the stringency of wartime requirements during the Second World War that stimulated the development of ergonomics as a systematic, integrated and essential discipline.

Servicemen had to operate in environmental conditions, both natural and artificial, that affected their efficiency and exposed them to hazards, which produced casualties not attributable to enemy action. Similarly, the rapid developments in technology resulted in weapons in which the performance of the man-machine system was clearly limited by the man and not by the machine. For the first time, the engineers and administrators, who had hitherto had a monopoly on all task designs, were forced to take serious account of the limitations of the human operator. The anatomists, physiologists and psychologists, who had previously worked only in universities and hospitals, were called on to advise on the limits of performance and the expected level of efficiency under given conditions of environmental stress.

The success of these efforts led to proposals after the war that these activities should be integrated and applied to industrial problems. One important step in this direction was the formation of the Ergonomics Research Society in Great Britain in 1949. There are now many countries with national ergonomics societies. The same activity is called " human engineering " in the USA, where more recently it has also been given such designations as " human factors " and " biotechnology ". There has been a tendency in the USA to isolate those aspects of work design that come within the province of the psychologist, describing them in such terms as " engineering psychology " and " applied experimental psychology ".

Because of the basic differences in concepts and the traditional academic separation, it took many years, in fact, before the anatomists, physiologists and psychologists really understood each other's problems and techniques. As a committed technologist interested in real problems and in actually solving them, the ergonomist has also to learn to communicate with other specialists such as design engineers, production engineers, work study specialists, cyberneticians, experts in operational research, environmental engineers, industrial hygienists and industrial medical officers, in addition to the great variety of policy-makers and decision-makers, such as managers, accountants, civil servants and military officers.

From all this activity it has gradually become apparent that the classical academic approach, based on pure disciplines and high standards of specialized evidence, is not enough. Questions of cost, speed, expediency and compromise present themselves and have to be dealt with. In the military field, as well as in the industrial, the ultimate criterion of cost effectiveness has been established. It is considerations such as these that govern the choice between alternative solutions. For example, with regard to the question of protecting a worker from a hostile climate by special clothing, there is often the alternative of removing that climate by the use of a special air-conditioned environment. Similarly, the need for effective interface design to reduce the level of skill required can be obviated by the design of better training schemes by which workers acquire the necessary skill more readily. If a work place has been so designed that its suitability is restricted to a specific group of the population, then sometimes it is preferable to provide techniques for selecting the appropriate part of the population than to redesign the work place.

It may be objected that all this takes no account of the fact that operators are human beings, individuals with a value and a dignity that, unlike material processes, cannot be measured merely in terms of efficiency and profits. Obviously this is true, but in practice it often proves faster and simpler to persuade the decision-makers to take proper account of the workers, not because of ethical considerations, but because it is either uneconomical or illegal not to do so. Fortunately it is usually true that, if fully comprehensive measures are applied, the interests of the workers as people coincide with the managers' interest in them as sources of productivity. In instances where this is not the case, legislation may provide the proper incentive.

In the years immediately after the Second World War ergonomics made little progress because of the uneconomic nature of many of the recommendations. This was largely because these recommendations essentially involved changes in already existing situations. The solution to this problem is to make the appropriate recommendations before the system is created in the first place. For example, it is much more economical to design equipment protecting workers from heat if the problem is raised before the hardware is built than to design such protection as an addition and an afterthought. Similarly, a well-designed dial is no more expensive to produce than a badly designed one if the ergonomist's recommendations are made and taken into consideration at the design stage.

This principle of introducing ergonomics at the right point in the design process is the essence of what is called " systems ergonomics ", in contrast to " classical ergonomics ", which is the piecemeal approach to environmental design. Systems ergonomics requires a much higher level of expertise. It is much easier to see what is wrong when a situation already exists than to predict, at an early stage in the design process, what ergonomics problems are likely to arise. However, the latter approach does make it possible to use much more fundamental thinking with correspondingly greater benefit. The classical ergonomics approach is to find out how a machine or situation can be most easily modified to suit the worker. The systems ergonomist asks the question: " Given that there is a total system problem that will inevitably be solved partly by the activities of men and partly by the activities of machines, which of the total activities are best allocated to men and which to machines? " This question of " allocation of function " between man and machine is governed by four criteria:

(1) relative abilities: this demands a knowledge in relative terms of which tasks men are basically suited for and in which tasks machines are better employed;

(2) cost/value: the relative cost in terms of acquisition, selection, training, subsistence and insurance for men *versus* capital investment and depreciation for machines;

(3) the integrated task: this involves the recognition that men are only available as integral units, and given that a man must be present for certain purposes he may as well do certain other tasks that could be done

by machines were it not for the availability of the man. In addition, the man as a human being requires a task that is integrated in the sense that it is worth doing and makes use of his abilities;

(4) graded tasks: in large systems it is not desirable that all tasks should be of equal difficulty. Members of the available working population always vary in their levels of ability, skill and seniority, and a good design will make use of this variety.

Only when this allocation of function has been completed in a systematic fashion is it necessary to put into practice the classical ergonomics procedures of matching men, machines, work spaces and environments.

One great advantage of the systems approach is that it is time-saving, since the personnel aspects of the system can be developed concurrently with the hardware aspects. However, this sort of advanced ergonomics can only be practised by the fully trained professional ergonomist. It is valuable and indeed now almost essential for the development of large-scale expensive systems, as well as for relatively simple, cheap systems where mass production is the aim.

The foregoing comments are not meant to imply that all ergonomics work should be done at this level and on such a scale. The majority of activities in this field are much more mundane, since a piecemeal approach must be adopted in dealing with a whole range of problems encountered in industry, agriculture, health, commerce, sport and every other sphere of human endeavour. Much of this work will not be done by full-time ergonomists, but by other specialists, whose responsibilities overlap with ergonomics. These include designers, industrial engineers and occupational health experts. The best service that the ergonomist can render his colleagues in these fields is to integrate the knowledge and techniques for a limited area, such as anthropometry, work-space layout, and dial design. This expertise can then be used by specialists practising in other fields who find it necessary or desirable to extend their activities into some aspect of ergonomics. There are obvious problems inherent in attempting to place complex tools in relatively inexperienced hands, particularly since, at present, ergonomics practice depends on extensive background knowledge that is not easily codified. Nevertheless, the attempt must be made, and one point in favour of this strategy is that much ergonomics practice does follow logically from a basic awareness of the limitations of people. It looks as though progress in ergonomics will continue to depend on this dual activity: highly specialized work by trained ergonomists and wide-ranging ergonomics practice by specialists in other fields.

By whatever name the discipline is known, its principles and practice are bound to endure, since no one can seriously dispute the basic premise, which is that the most important factor in work design is the worker.

RETROSPECT AND PROSPECT

A. SCIENCES AND TECHNOLOGIES RELATED TO ERGONOMICS

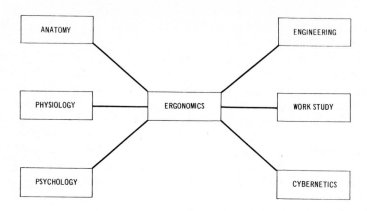

B. HISTORICAL DEVELOPMENT OF ERGONOMICS

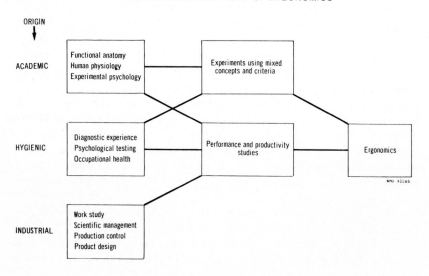

C. DESIGN PROCEDURE IN SYSTEMS ERGONOMICS

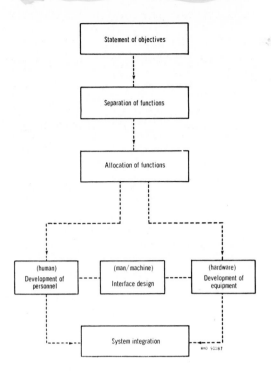

D. RELATIVE ADVANTAGES OF MEN AND MACHINES

Characteristic	Machine	Man
Speed	Much faster.	Quickest reaction time 0.05 second.
Power	Consistent at any level. Large, constant standard forces.	2.0 hp for about 10 seconds 0.5 hp for a few minutes 0.2 hp for continuous work over a day.
Consistency	Ideal for: routine; repetition; precision.	Not reliable: should be monitored by machine.
Complex activities	Multi-channel	Single-channel.
Memory	Best for literal reproduction and short term storage.	Large store, multiple access. Better for principles and strategies.
Reasoning	Good deductive.	Good inductive.
Computation	Fast, accurate. Poor at error correction.	Slow, subject to error. Good at error correction.
Input sensitivity	Some outside human senses, e.g., radioactivity Can be designed to be insensitive to extraneous stimuli.	Wide energy range (10^{12}) and variety of stimuli dealt with by one unit; e.g., eye deals with relative location, movement and colour. Good at pattern detection. Can detect signals against high levels of background noise. Affected by heat, cold, noise and vibration (exceeding known limits).
Overload reliability	Sudden breakdown.	Gradual degradation.
Intelligence	None.	Can deal with unpredicted and unpredictable; can anticipate.
Manipulative abilities	Specific.	Great versatility.

REVIEWERS

Dr Harwood S. Belding, Department of Occupational Health, Graduate School of Public Health, University of Pittsburgh, Pa., USA

Dr F. H. Bonjer, Head, Dutch Institute of Preventive Medicine, Leiden, Netherlands

Dr S. K. Chatterjee, Medical Deputy Director, Central Labour Institute, Sion, Bombay, India

Dr F. N. Dukes-Dobos, Environmental Physiologist, Bureau of Occupational Safety and Health, Department of Health, Education, and Welfare, Cincinnati, Ohio, USA

Professor E. Grandjean, Department of Hygiene and Applied Physiology, Swiss Federal Institute of Technology, Zurich, Switzerland

Professor M. J. Karvonen, Director, Institute of Occupational Health, Helsinki, Finland

Dr K. Kogi, Chief, Laboratory of Work Physiology, Institute for Science of Labour, Setagaya-ku, Tokyo, Japan

Professor F. Lavenne, Head, Department of Medicine, Catholic University of Louvain, Belgium

Dr C. S. Leithead, Professor of Medicine, Department of Medicine, Haile Sellassie I University, Addis Ababa, Ethiopia

Professor N. Lundgren, Chief, Department of Work Physiology, National Institute of Occupational Health, Stockholm, Sweden

Professor R. A. McFarland, Harvard School of Public Health, Boston, Mass., USA

Professor B. Metz, Centre for Bioclimatic Studies (C.N.R.S.), Strasbourg, France

Professor I. V. Muravov, Director, Scientific Research Institute for Medical Problems of Physical Culture, Kiev, USSR

Professor C. P. Odescalchi, Via Conservatorio 30, Milan, Italy

Dr N. Pardon, Consultant Physician, Centre for Information on Industrial Medical Services, Paris, France

Professor B. Shackel, Department of Ergonomics and Cybernetics, University of Technology, Loughborough, England

Dr H. G. Wenzel, Head, Second Physiological Department, Max Planck Institute of Occupational Physiology, Dortmund, Federal Republic of Germany

Professor A. Wisner, Laboratory of Work Physiology, National Conservatory of Arts and Crafts, Paris, France